簡材料 · 滋味餸

Gourmet Cooking with Simple Food

U0111346

代序

每一個人心中都有一位大師！
廖師傅是很多人心中的大師傅！

從第一次上師傅的課，會被師傅教學的風範吸引着，很多大家從來不知道的入廚知識，師傅都會透過每道菜式，深入淺出，以精采的演繹、一堆堆素材，瞬眼變成一道道色香味俱全的菜餚。

沒有甚麼材料可以難到一位全能的師傅，那怕是最普通的簡單材料。比喻大排檔的菜式，師傅把一些簡簡單單、在街市隨手買到的基本食材，只要發揮得宜，可煮出家常戶曉的大排檔菜式，不單易上手，味道比我們在大排檔吃到的，更勝一籌！

不管你入廚經驗有多少，不管你煮的是珍饈百味，還是家庭小菜，只要選材得宜，配合適當的技巧，師傅總有法子教大家做得一手出色菜餚，令大家成為當家大廚！

Rachel Yau
Rachel's Kitchen 主理人

認識廖教賢師傅已經 20 多年，感覺他一直都是那麼敬業樂業、風趣健談，所以無論是當年的酒家掌廚到專業烹飪導師、電視節目廚藝專家及食譜作者，每一個角色的精采演繹，都深受歡迎！

廖師傅有着豐富的教學經驗與廣博的廚藝知識，每每總會站在不同人的角度，思其所需，並對症下藥地為喜歡烹飪的朋友，設計撰寫一系列貼切實用且美味的食譜。

今次新作品《簡材料 · 滋味餸》一定會為大家帶來更多烹調的新啟發，並感受如何發揮簡單材料的巧妙廚藝。

陳永瀚師傅
烹飪班導師、食譜書作者

師傅又出書啦！

廖教賢老師一向深得同學愛戴，也是受讀者們喜愛的一位烹飪導師。

我們還記得在職訓局的酒店系就讀中餐課程的日子，我們這班學生是廖師傅執教的第一班學生。當年我們十分頑皮，無心向學，甚麼餐飲理論、飲食文化、八大菜系等……那有興趣上心，但到師傅教授的廚務實習課堂，我們一定是最開心快樂，不但得到師傅的悉心教授，又可自己大展身手，最後還有一頓豐富的午餐或晚餐，此情此景，歷歷在目。轉眼已差不多二十個年頭，師傅也退休了好幾年，但同學們仍期待上廖師傅的課堂呢！前幾年曾與師傅合作編寫食譜，過程中獲益不少，也是我的榮幸。

原來，退休只是師傅人生中另一個高峰的開始！退休後的他，忙過不停，在各方平台教授廚藝，他的著作一本比一本精采實用，今次的新作必定令讀者們帶來無限驚喜！

在此，我多謝在我人生中最重要的人，他不但教曉我廚藝知識，還有人生道理，使我在人生路上獲益良多，多謝您～廖師傅！

師傅，希望您保重身體，與師母繼續享受人生。
最後，預祝師傅新書一紙風行！

林勤樂師傅

酒店主廚、食譜書作者、電視節目廚藝專家

前言

廖師傅
談出版緣起

天天做飯，人人都想好吃又有心思，這個很容易。
過往我的作品以不同的方向為主題，大多離不開以不複雜的烹調法及簡單廚具為大前提。編寫方向以讀者的烹飪角度為出發點，我的菜譜目的也如此，盡量以簡單易明的演繹方法編寫，易明易做，誰都可發揮自己的廚藝，務求讀者在自家廚房盡情發揮，在追求美味菜餚之餘，也可掌握色、香、味、形的要素。

讀者對坊間很多食譜，都抱着嘗試烹煮的態度，無奈地當中有些材料難以選購、不輕易找到。他們一直疑惑：「如果不用這款材料，有其他材料代替嗎？」這次新書以「簡材料」為主題，在一般街市購買得到的材料為大家烹煮美味菜餚。書中以食材品種分類，分為海產、肉類、豆品、蔬菜及稻米五大類別，另外還增加備貨篇，使讀者在烹調前作好準備，下廚更方便。

希望這本新作令讀者更容易發揮其烹飪創意空間，透過文字與圖片，一看即會，一試可行。「簡而備，易而行」，為家人摯愛煮出美味菜餚，與眾同樂～是我的目標。

在此特別鳴謝 Funny Oasis 及 Joey 全力支持及協助拍攝，更多謝萬里機構的製作團隊用心製作，令新作品更完美地與大家見面。

目錄

金華火腿茸

材料

金華火腿肉 200 克
清水 1 杯

做法

1/ 金華火腿肉洗淨，飛水，加入清水約 1 杯浸過面。
2/ 放入鑊蒸約 40 分鐘，取出待涼，剁成火腿茸，放入密封玻璃樽，貯存雪櫃約可待一個月。

Chef's Tips

- 金華火腿茸用途廣泛，如湯羹、蒸製菜式裝飾之用。
- 金華火腿用水浸着蒸，較容易軟腍。
- 蒸金華火腿的湯汁可作上湯使用。
- 先用刀略剁鬆肉質，剁碎時才不會四處飛散。
- 如熬製上湯時，可將金華火腿湯渣製成火腿茸。

Finely Chopped Jinhua Ham

Ingredients

- 200 g Jinhua ham, 1 cup water

Method

1. Rinse the Jinhua ham and scald. Cover with 1 cup of water in a bowl.
2. Steam in a wok for about 40 minutes, remove and finely chop when cool, store in a sealed glass jar. Keep in a refrigerator can last for one month.

Chef's Tips

- Jinhua ham is a brilliant garnishment widely used in soups or steamed dishes.
- Jinhua ham is easier to become soft by steaming it covered with water.
- The sauce from the steamed Jinhua ham can be used as stock.
- Chop the Jinhua ham gently to loosen the meat texture, it will not splash everywhere when finely chopping.
- If you are cooking stock with Jinhua ham, you may use the ham leftover and finely chop.

上湯
Stock

材料

光雞 1 隻，瘦豬肉 2.5 斤，金華火腿肉 250 克，金華火腿骨 100 克，白胡椒粒 1 茶匙，圓肉 20 克，大薑片 50 克，清水 5 公升

做法

1/ 光雞洗淨；豬肉洗淨，切件，所有肉料飛水，沖洗。
2/ 大煲內注入清水 5 公升，放入所有材料，用大火煮滾（撇去血沫），轉小火煲 5 至 6 小時，關火。
3/ 撇去浮面的油分，盛起湯渣，用密孔篩網或紗布隔去肉沫，待涼，貯存雪櫃可使用數天。

Ingredients

1 chicken, 1.5 kg lean pork, 250 g Jinhua ham, 100 g Jinhua ham bone, 1 tsp white peppercorns, 20 g dried longans, 50 g large ginger slices, 5 litres water

Method

1. Rinse the chicken and lean pork. Cut the pork into pieces. Scald and rinse.
2. In a large pot, pour in 5 litres of water. Add all the ingredients, bring to the boil over high heat (skim off any froth), turn to low heat and simmer for 5 to 6 hours. Turn off the heat.
3. Skim off any oil on the surface, take out the residues, strain the stock with a meshed strainer or a piece of gauze cloth to form clear chicken stock. When cool, refrigerate for use in a couple of days.

Chef's Tips

- 用上湯烹調湯羹，令味道鮮味倍增。
- 金華火腿宜飛水 5 分鐘，可去掉油分，令上湯味道更佳，可使用連皮金華火腿肉。
- 如不想油分太多，先將光雞去皮才煲湯。
- 切忌使用大火，會令上湯渾濁不清。

Chef's Tips

- Cooking the soup with the stock, it is doubly flavourful.
- The Jinhua ham needs to scald at least 5 minutes for removing grease, making the stock more delicious. You can use the Jinhua ham with the skin.
- If you want a clearer stock, remove the chicken skin first.
- The stock will be not clear if cooking over high heat.

墨魚膠
Cuttlefish Paste

詳細做法睇片！

Ingredients

- 500 g cuttlefish meat, 12 skinned water chestnuts

Seasoning

- 1/2 tsp salt, 1/2 tsp chicken bouillon powder, 2 tbsp corn flour, 1/3 tsp ground white pepper, 1/2 tsp sesame oil

Method

1. Rinse the cuttlefish meat, remove the skin, wipe dry and cut into small pieces. Mince the meat in a food processor, put into a stainless steel pot.
2. Bash the water chestnuts with a knife and finely chop.
3. Knead the cuttlefish paste heavily, stir in one direction until sticky. Add the seasoning, stir in one direction again until stickier, throw into the pot for a few times.

Chef's Tips

- You can make deep-fried cuttlefish balls, cuttlefish ball congee, fried cuttlefish cake and stuffed trio.
- Both fresh and chilled cuttlefish are available in the market. If you use fresh cuttlefish, better cut off the suckers along their tentacles before blending; or the granules will spoil the meat texture.
- The cuttlefish paste must be stirred in one direction, or the paste will not taste spongy.
- If you are not prepare to use the cuttlefish paste right after it is made, refrigerate for 1 to 2 hours for a better result.
- You can add 50 g of diced fat pork to the paste after finishing the kneading steps. The oil will be come out if kneading too much, which will give the paste a loose texture.

材料

墨魚肉 500 克，馬蹄肉 12 粒

調味料

鹽半茶匙，雞粉半茶匙，粟粉 2 湯匙，胡椒粉 1/3 茶匙，麻油半茶匙

做法

1/ 墨魚肉洗淨，去衣及抹乾，切小件，放入攪拌機攪成墨魚膠，置於不銹鋼窩。
2/ 馬蹄肉用刀拍碎，剁成碎粒。
3/ 用力擦墨魚膠十數下，順一方向攪至黏稠，加入調味料順方向拌至黏力增強，撻十數下即成。

Chef's Tips

- 可製成炸墨魚丸、生滾墨魚丸粥、香煎墨魚餅、煎釀三寶等。
- 墨魚分為鮮貨及冰鮮貨，如用鮮貨建議切掉觸鬚的吸盤，否則攪碎後混有硬粒，影響質感。
- 必需順一方向攪動墨魚膠，否則墨魚膠鬆散、欠彈力。
- 墨魚膠如不立即使用，放入雪櫃冷藏 1 至 2 小時，效果更佳。
- 可依傳統做法加入肥肉粒 50 克，甘香美味。待墨魚膠完成後才放入肥肉粒攪拌，否則過度搓擦易滲出油分，令墨魚膠鬆散。

豬肉丸
Pork Balls

材料

梅頭瘦肉 400 克（攪碎），薑 40 克（切幼粒），蝦米 60 克

調味料

鹽 2/3 茶匙，雞粉半茶匙，粟粉 30 克，冷水 65 毫升，鹼水 1/4 茶匙，麻油半茶匙，胡椒粉 1/3 茶匙

做法

1/ 蝦米用水浸軟，切幼粒，用白鑊以小火慢慢炒香。
2/ 將調味料的冷水與鹼水混合；豬肉碎與其他調味料順一方向攪動，水分逐少加入，起膠後撻撻十數下，冷藏 2 小時。
3/ 加入蝦米及薑粒拌至起膠，撻撻，冷藏 1 小時，唧成丸狀隨時使用。

Chef's Tips

- 可製成生滾肉丸粥、蘿蔔絲蒸肉丸等菜式。
- 鹼水在雜貨店或糕點食材專門店有售，有大瓶裝或小份量出售。
- 鹼水有吸收水分的作用，但不可直接加入，宜先用水稀釋。

Ingredients

400 g pork collar-butt (minced), 40 g ginger (finely diced), 60 g dried shrimps

Seasoning

2/3 tsp salt, 1/2 tsp chicken bouillon powder, 30 g corn flour, 65 ml cold water, 1/4 tsp food-grade lye, 1/2 tsp sesame oil, 1/3 tsp ground white pepper

Method

1. Soak the dried shrimps in water until soft and finely dice. Stir-fry in a wok without oil over low heat until aromatic.
2. Combine the cold water with food-grade lye of the seasoning. Mix the minced pork with the other seasoning and stir in one direction. Add the lye water little by little until gluey. Knead and throw for a few times and refrigerate for 2 hours.
3. Add the dried shrimps and ginger and stir until sticky. Throw into the bowl and refrigerate for 1 hour. Squeeze into balls for use anytime.

Chef's Tips

- You can make pork ball congee, steamed pork balls with radish, etc.
- Food-grade lye is available in a bottle or small quantities in grocery stores or specialty shops for cake ingredients.
- The purpose of adding food-grade lye is to absorb water, it needs to be diluted with water beforehand.

醃味鹹酸菜
Pickled Mustard Greens with Additional Seasoning

材料

鹹酸菜 1 斤，鹽 3 茶匙

糖醋水

白米醋 200 毫升，清水 150 毫升，砂糖 300 克，酸梅 2 粒，紅尖椒 1 隻（開邊）

做法

1/ 糖醋水用慢火煮溶（紅尖椒除外），待涼，放入紅尖椒浸泡，備用。
2/ 鹹酸菜剪去葉，留梗放入大窩，灑上鹽拌勻醃約 8 至 10 分鐘，沖洗鹽分。
3/ 鹹酸菜浸水約 3 至 4 小時，期間多次換水去除鹹味。
4/ 鹹酸菜梗切片，放入糖醋水浸漬，貯存雪櫃一天後使用。

Ingredients

600 g pickled mustard greens, 3 tsp salt

Sweet Vinegar Water

200 ml white rice vinegar, 150 ml water, 300 g sugar, 2 pickled plums, 1 red chilli (cut into half)

Method

1. Cook the ingredients of sweet vinegar water (except the red chilli) over low heat until dissolve. When cool, soak in the red chilli.
2. Cut away the leaves of the pickled mustard greens, put the stems into a big pot. Sprinkle with salt and mix well. Marinate for 8 to 10 minutes, and rinse to remove salt.
3. Soak the pickled mustard greens in water for 3 to 4 hours, change water repeatedly to reduce salty taste.
4. Slice the stems of the pickled mustard greens, pickle in the sweet vinegar water, chill for one day before use.

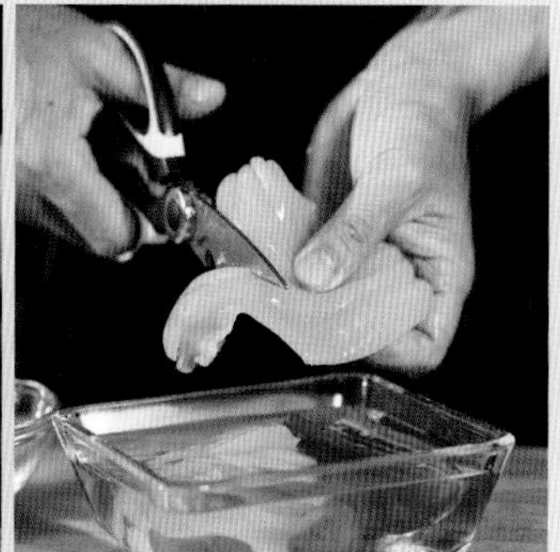

Chef's Tips

- 自行醃漬鹹酸菜，帶出鹹、甜、酸多層次的味道，可製成味菜炒肉絲、炒鵝腸等。
- 浸漬鹹酸菜越久越好，令糖酸味更易突出。
- 剪出來的鹹酸菜葉可煲湯，不要棄掉。

Chef's Tips

- We can pickle pickled mustard greens in sweet vinegar water at home to give a multiple levels of salty, sweet and sour flavour. They can be stir-fried with pork or goose intestine.
- Pickle the pickled mustard greens for a period as long as possible. The longer the time, the more outstanding is the sweet sour flavour.
- Do not waste the leaves cut out from the pickled mustard greens. They can be used for cooking soup.

馬拉盞
Belachan

材料

蝦醬 1 樽（約 280 克），蒜子 10 粒，乾葱頭 6 粒，指天椒 2 隻，香茅 1 枝，芫茜 3 棵，蝦米 40 克，砂糖 5 茶匙，花雕酒 6 湯匙，生油 30 毫升

做法

1/ 蒜子、乾葱頭、指天椒、香茅及芫茜放入攪拌機，加入清水 1 杯攪碎，隔水、留渣備用。
2/ 蝦米用水浸軟，隔水，略剁碎。
3/ 燒熱鑊下生油，轉小火，放入攪碎料炒香，下蝦米再炒香，拌入蝦醬及花雕酒慢慢煮熱，用小火煮約 5 分鐘。
4/ 加入砂糖煮溶，待涼，放入密封玻璃樽。

Chef's Tips

- 可烹調馬拉盞炒通菜、馬拉盞炒飯等，惹味好吃。
- 不同品牌的蝦醬稀稠度有別，如太稠的蝦醬可酌加花雕酒煮稀。
- 不要加水炒蝦醬，否則貯存時易變壞。

Ingredients

1 bottle shrimp paste (about 280 g), 10 cloves garlic, 6 shallots, 2 bird's eye chillies, 1 stalk lemongrass, 3 stalks coriander, 40 g dried shrimps, 5 tsp sugar, 6 tbsp Shaoxing wine (Hua Diao), 30 ml oil

Method

1. Blend the garlic, shallots, bird's eye chillies, lemongrass and coriander with 1 cup of water in a food processor. Strain the sauce and keep the crushed ingredients.
2. Soak dried shrimps in water until soft, drain and roughly chop.
3. Heat a wok, add some oil, turn to low heat and stir-fry the crushed ingredients until fragrant. Add the dried shrimps and stir-fry until aromatic. Mix in the shrimp paste and Shaoxing wine, heat up slowly, simmer over low heat for about 5 minutes.
4. Add the sugar and cook until dissolves. When cool, store in a sealed glass jar for use later.

Chef's Tips

- Belachan can be stir-fried with water spinach, fried with rice, etc.
- The consistency of shrimp paste varies with different brands. If it is too thick, dilute with Shaoxing wine while cooking.
- Do not cook the shrimp paste with water, it will easily spoil in storage.

黑椒醬汁
Black Pepper Sauce

材料

黑胡椒碎粒 40 克，香葉 4 片，草果 2 粒，清水 2 杯

調味料（拌勻）

鹽 1/3 茶匙，蠔油 1 湯匙，砂糖 2 茶匙，老抽 2 茶匙，粟粉 2 茶匙，開水 2 湯匙

做法

1/ 所有材料蒸約 1 小時，隔去材料（取黑胡椒碎 2 茶匙備用），黑椒汁留用。
2/ 煮熱黑椒汁，加入調味料煮滾成黑椒醬汁，下預留之黑胡椒碎拌勻，待涼，放入密封玻璃樽貯存。

Chef's Tips

- 用黑椒汁可炒牛柳絲、牛柳粒或煎牛仔骨等菜式。
- 香葉及草果在雜貨店或泰國食材店有售。
- 如不喜歡醬汁帶有黑胡椒碎粒，可不添加，悉隨尊便。

Ingredients

40 g crushed black peppercorns, 4 bay leaves, 2 nutmegs, 2 cups water

Seasoning (mixed well)

1/3 tsp salt, 1 tbsp oyster sauce, 2 tsp sugar, 2 tsp dark soy sauce, 2 tsp corn flour, 2 tbsp drinking water

Method

1. Steam all the ingredients for about 1 hour. Strain the ingredients (reserve 2 tsp of black peppercorns) and keep the black pepper sauce.
2. Heat the black pepper sauce, add the seasoning, bring to the boil, put in the black peppercorns and mix well. When cool, store in a sealed glass jar.

Chef's Tips

- You can make stir-fried beef tenderloin, fried beef short ribs with black pepper sauce.
- Bay leaves and nutmegs are sold at grocery stores or Thai food ingredient shops.
- You can skip the crushed black peppercorns in the sauce if you dislike.

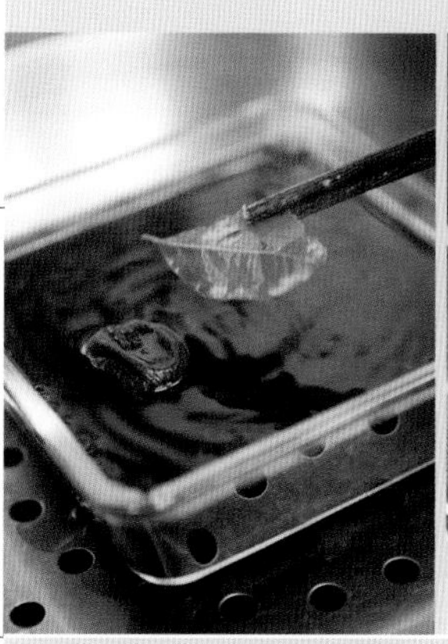

海
愛鮮甜
產

Seafood

啖啖嫩滑的龍躉魚肉，是貴賓式的招待

碧綠炒魚柳

材料

龍躉魚肉……300克
芥蘭……400克
香芹……80克
菜脯……40克

料頭

蒜茸……半茶匙
薑絲……10克
甘筍絲……10克

醃料

鹽……1/3茶匙
粟粉……2茶匙
蛋白……1湯匙（約半個）

調味料1（灼芥蘭）

鹽……2茶匙
砂糖……2茶匙
生油……2茶匙

調味料2（炒魚柳）

鹽……1/3茶匙
蠔油……1茶匙
麻油……1/3茶匙
胡椒粉……適量
粟粉……1茶匙
清水……2茶匙

做法

1/ 芥蘭切掉菜葉，留梗切成小段約 5 厘米長，用小刀在菜梗兩端縱橫剔入 1.5 厘米深，浸水約半小時。

2/ 魚肉切厚片，再切成幼條狀，將醃料拌勻，倒入魚柳內混和醃味。

3/ 香芹摘葉及切小段，洗淨。菜脯切小段，再切成幼條，用熱水浸泡令味道不太鹹。

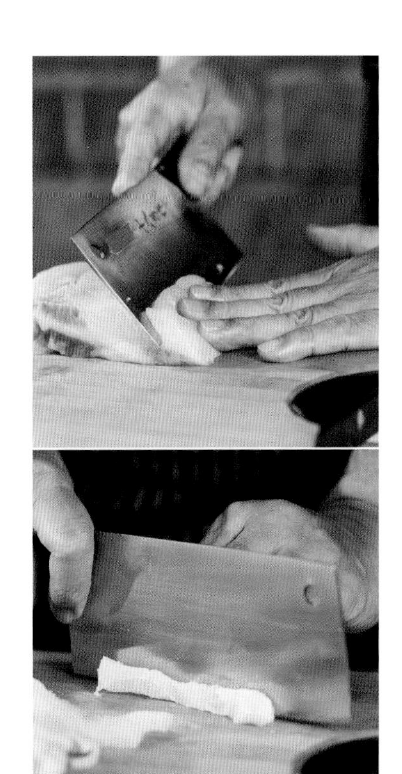

4/ 鑊中放入清水 2 杯燒滾，加入調味料 1 及芥蘭灼約 2 分鐘，下香芹及菜脯稍灼，隔起水分。

5/ 鑊燒熱後加入生油約半杯，用中油溫燒熱，放入魚柳泡油至熟，盛起隔油。

6/ 鑊內放入料頭爆香，加入芥蘭略炒，再放入魚柳，調味料 2 混合後慢慢倒入，用中火與其他材料炒勻，上碟享用。

Chef's Tips

- 若芥蘭梗不夠嫩，可稍削表皮較厚端部分。當兩端剔開後浸水，時間不宜太久，因泡水後芥蘭會收縮至過分捲曲，炒煮時較易折斷。

- 建議選用刀刃薄的小刀剔開芥蘭梗，否則容易爆開，影響賣相。

豉椒鹹菜炒鮮魷

酸、甜、微辣，是最惹味的下飯菜

材料

鮮魷魚……300 克
青甜椒……1 個
紅尖椒……1 隻
洋葱……半個
已醃鹹酸菜……200 克
（做法參考 p.13）

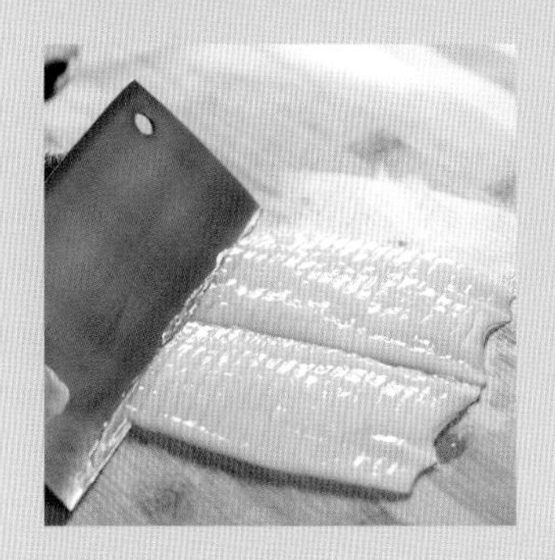

料頭

蒜茸……半茶匙
薑……6 片
豆豉……2 茶匙

調味料

鹽……1/3 茶匙
蠔油……2 茶匙
老抽……1 茶匙
砂糖……1 茶匙
粟粉……1 茶匙
清水……1 湯匙
紹酒……1 茶匙（後下）

Chef's Tips

- 鮮魷剞花紋時，建議在腹腔面下刀，較容易切出花紋，但不要剞得太深，以免切斷鮮魷。
- 鮮魷魚的翼位較韌，可切掉；魷魚的薄位則切成條狀。
- 灼鮮魷時以免過熟，令收縮過火口感變老。
- 鹹酸菜經處理醃製後，加添了一份酸甜味道，美味升級。

做法

1/ 青紅椒去籽，與洋葱同切角。鹹酸菜梗斜刀切片，用乾鑊炒至乾身及香。
2/ 鮮魷撕去表面薄膜，剪開腹腔洗淨，縱橫剞花紋，切成長三角形小件。
3/ 鮮魷件放入滾水灼熟，瀝乾水分。
4/ 鑊內下生油 1 湯匙，放入料頭爆香，加入青紅椒、洋葱及清水 2 湯匙，用中火炒熟。
5/ 放入鮮魷及鹹酸菜略炒，加入調味料用猛火炒勻，最後灒入紹酒 1 茶匙略炒即可上碟。

嘗一啖，鮮、嫩、滑，你會歡呼大叫

豆苗魚茸羹

材料

桂花魚......2條（每條約1斤）
豆苗......400克
淡滑豆腐......1磚
甘筍......80克
金華火腿肉......50克
雞蛋......1個
上湯......3杯（參考p.10）

調味料

鹽......半茶匙
胡椒粉......1/4茶匙
麻油......1/3茶匙

粟粉水

粟粉......2湯匙
清水......4湯匙

做法

1/ 桂花魚洗淨，放於碟上，排上薑片在魚面蒸約12分鐘至熟，稍涼，將魚肉拆出，小心去掉魚骨，魚肉弄散成魚茸。

2/ 豆苗洗淨，剁碎；鑊中下生油1湯匙，放入豆苗及鹽1茶匙，用中火將豆苗炒熟，壓去水分。

3/ 甘筍切薄片；滑豆腐切小粒。金華火腿肉用水浸過面蒸約20分鐘，取出剁成火腿茸，火腿湯可加入上湯使用。

4/ 鑊中放入上湯煮滾，加入魚茸、豆苗及甘筍片煮滾，加入調味料及豆腐粒，用粟粉水勾芡，煮滾後倒入蛋液拌成蛋花，盛起，灑上火腿茸即可享用。

Chef's Tips

- 在挑選魚骨時要反覆檢查數次，以防小魚骨留在魚茸內。
- 豆苗要盡量壓去水分，以免釋出水分，令湯味變淡。
- 豆腐粒在最後步驟才放入湯內煮熱，否則容易煮爛及過老。

惹味的胡椒蟹，食指大動

金不換胡椒焗蟹

材料

肉蟹……1隻
金不換……2棵
鮮青胡椒……2串
白胡椒粒……1茶匙
紅尖椒……1隻
牛油……40克

調味料

鹽……半茶匙
砂糖……半茶匙
蠔油……1湯匙
紹酒……1湯匙
清水……1杯

料頭

蒜……3粒（切片）
薑……8片
紅葱頭……4粒（切片）

做法

1/ 金不換摘葉留用；白胡椒粒用刀略拍成粗粒；鮮青胡椒串洗淨；紅尖椒切片。

2/ 肉蟹在腹部中央用刀破開，拉開蟹蓋，取去內臟洗淨，斬件，蟹鉗用刀拍裂，灑上粟粉。

3/ 鑊中下生油燒熱，放入蟹件泡油至熟，盛起隔油。

4/ 煮溶牛油，加入料頭及白胡椒粒炒香，下蟹件略炒，倒入調味料轉慢火，加入鮮青胡椒串，加蓋焗約 6 分鐘，下金不換葉及紅尖椒炒勻，略焗 1 分鐘炒乾，上碟。

Chef's Tips

- 蟹件泡油時，不宜放入蟹蓋，因蟹膏容易加熱過老，影響質感，只要煮時與蟹件一同加蓋焗 6 分鐘即可。
- 肉蟹大小有別，如選用較大的肉蟹，蟹鉗較難熟透，烹調時間可能要增加約 2 至 3 分鐘，水分也要略添加。

飄飄椰香，清新、美味

椰香蛋白蒸斑球

材料

石斑肉……500 克
椰皇……2 個
蛋白……4 個
葱……1 棵（切絲）
紅尖椒……1 隻（切絲）

調味料 1（蛋白）

蛋白……150 克
椰皇水……150 毫升
鹽……1/3 茶匙
粟粉……1 茶匙
胡椒粉……適量

調味料 2（石斑肉）

鹽……1/3 茶匙
雞粉……1/4 茶匙
粟粉……2 茶匙
蛋白……2 茶匙
麻油……1 茶匙
胡椒粉……1/4 茶匙
紹酒……2 茶匙
蒜茸……2 茶匙
幼薑粒……2 茶匙

做法

1/ 椰皇破開，盛起椰皇水，椰肉取出切幼粒，調味料 1 混和後，加入椰肉拌勻。

2/ 石斑肉切長方形厚片（約 4 × 2 厘米），洗淨，瀝乾水分，加入調味料 2 醃約 10 分鐘。

3/ 石斑肉放入窩碟內鋪平，蒸約 4 分鐘，倒去水分，加入步驟 1 的調味料再蒸 4 分鐘至熟，灑上葱絲及紅椒絲即可。

Chef's Tips

- 將蛋白調味料混和時，最好先將椰皇水與粟粉開溶，才加入其他材料，以免粟粉溶解不勻。
- 倒入蛋白調味料後，最宜用小火蒸熟，否則蛋白容易過老。

蔥味椒蒜香，滋味無窮

勁蒜剁椒蟶子皇

材料

蟶子皇..................4 隻

蔥........1 條（切絲）

芫茜........................1 棵

紅椒絲.................適量

料頭

蒜頭...8 粒（剁茸）

紅尖椒..................2 隻

薑片....................30 克

醬汁料

生抽.................3 湯匙

蠔油.................1 湯匙

麻油.................2 茶匙

香醋.................1 湯匙

砂糖.................1 茶匙

清水.................2 湯匙

做法

1/ 將適量冰粒放入 4 杯開水中，備用。

2/ 蟶子皇洗淨，放入滾水灼約 20 秒，盛起，放入冰水浸涼。

3/ 蟶子肉抹乾水分，切約 3 至 4 小段，放回殼內，置於雪櫃稍冷藏。

4/ 料頭切幼粒，在鑊中下生油 1 茶匙用中小火爆香料頭，再加入醬汁料煮滾，盛起。

5/ 在雪櫃取出蟶子皇，淋上醬汁料，灑上蔥絲、紅椒絲及芫茜享用。

Chef's Tips

- 灼蟶子時火力要猛，但不要太久，以免過火令蟶子收縮至韌。

精緻的外型，未吃已令人心花怒放

花姿菊花盞

詳細做法睇片！

材料

墨魚膠............................200 克（做法參考 p.11）

方形春卷皮..6 片

蟹籽..30 克

做法

1/ 墨魚膠分成 12 小丸狀。

2/ 春卷皮對摺，剪成兩張分別 10 x 20 厘米長方形。

3/ 每張長方形再對摺，用剪刀在對摺處 45 斜度剪成條紋，深度是長方形的一半，而每條斜條約半厘米至 1 厘米寬。

4/ 將小墨魚丸放於長條春卷皮前端（沒斜紋部分），向內慢慢捲入成菊花形，用少許墨魚膠黏住封口，以牙籤固定。

5/ 鑊內下生油約 3 杯，用中火燒熱至中油溫（約 150℃），逐件放入油鑊炸熟至金黃色（約 3 分鐘），盛起，在菊花中央以少許蟹籽裝飾即成。

Chef's Tips

- 不宜釀入太多墨魚膠，會影響菊花盞的形態。
- 於春卷皮修剪斜紋時，以半厘米寬度較佳，因炸製成型時花瓣會更密、更美觀。
- 炸製時菊花盞的斜紋條建議盡量展開，較容易製成菊花形態。
- 蝦膠可代替墨魚膠，效果也相似。

黑胡椒濃、薄荷甜香，一次滿足兩種味道

黑椒薄荷蝦

材料

鮮蝦……12隻

三色甜椒…200克（去籽、切塊）

薄荷葉……50克

黑椒薄荷汁

黑胡椒幼粒……2茶匙

薄荷啫喱醬……2湯匙

新鮮薄荷葉……約10片（切碎）

蠔油……1湯匙

鹽……1/3茶匙

砂糖……1茶匙

清水……3湯匙

粟粉……1茶匙

醃料

鹽……1/3茶匙

粟粉……2茶匙

雞蛋液……3茶匙

麻油……半茶匙

做法

1/ 黑椒薄荷汁煮熱，備用。

2/ 薄荷葉洗淨，吸乾水分；三色甜椒洗淨，切角。

3/ 鮮蝦剝掉蝦身中間外殼，留頭尾，用刀在蝦背剟開，除去蝦腸，洗淨後用乾布吸去水分，用醃料拌勻。

4/ 鑊內下生油半杯燒熱至中油溫約140℃，下薄荷葉炸脆盛起；鮮蝦泡油至熟盛起。

5/ 炒鑊爆香三色甜椒，加入黑椒薄荷汁煮熱，倒入鮮蝦球炒勻上碟，灑上炸脆薄荷葉即可。

Chef's Tips

- 市面上出售的玻璃樽裝薄荷醬一般有兩款：薄荷醬汁——味道微酸而較稀；薄荷啫喱醬——味道微甜而較稠。
- 薄荷葉必須吸乾水分才下鑊炸脆，不宜高溫油炸，因葉片很薄容易焦燶。

金黃彈牙的魚丸，陣陣香茅味，引人食欲

香茅墨魚丸

材料

急凍墨魚肉............... 300 克

馬蹄肉.............................6 粒

香茅...................................6 支

麵包糠........................ 200 克

調味料

鹽................................. 半茶匙

雞粉.........................1/3 茶匙

粟粉.............................3 茶匙

麻油............................ 半茶匙

胡椒粉.....................1/4 茶匙

Seafood

做法

1/ 墨魚肉洗淨，抹乾水分，切小塊，放入攪拌機攪碎；馬蹄肉洗淨，用刀拍碎。

2/ 墨魚肉放入大碗內，用手搓擦數次，順一方向攪動約 1 分鐘，加入調味料拌勻，攪動至起膠及黏手，拿起搓撻數次，加入馬蹄肉拌勻成墨魚膠，用手唧成丸狀。

3/ 香茅洗淨，切掉尾部，留頭部較粗部分約 10 厘米長度，破開兩邊，手沾少許水將墨魚丸穿過香茅，黏上麵包糠。

4/ 生油用中火燒熱至中油溫，放入墨魚串炸至金黃色，上碟即可。

Chef's Tips

- 攪動墨魚膠時必須同一方向攪拌，向不同方向攪動，膠質會鬆散。
- 完成墨魚膠製作後，如非即時使用，最宜先放雪櫃冷藏約 2 小時或以上，令魚膠更有彈性。
- 將香茅破開兩邊釀入墨魚丸，讓香茅的香氣更易散發並滲入魚丸。

玉簪蝦球

爽脆的蝦肉、優雅的賣相，視覺與味覺大滿足

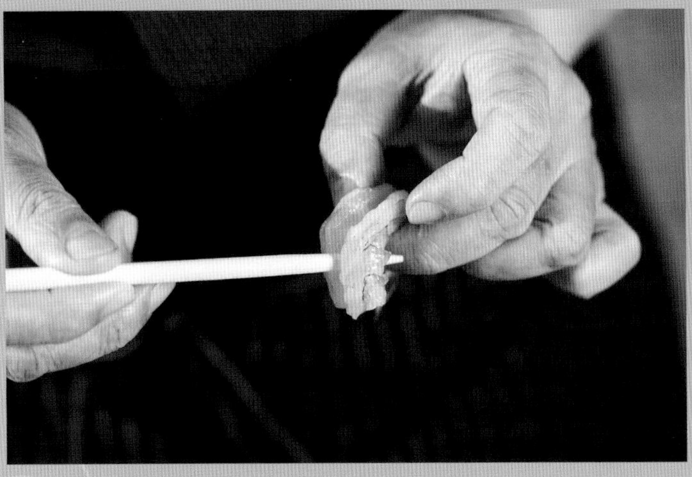
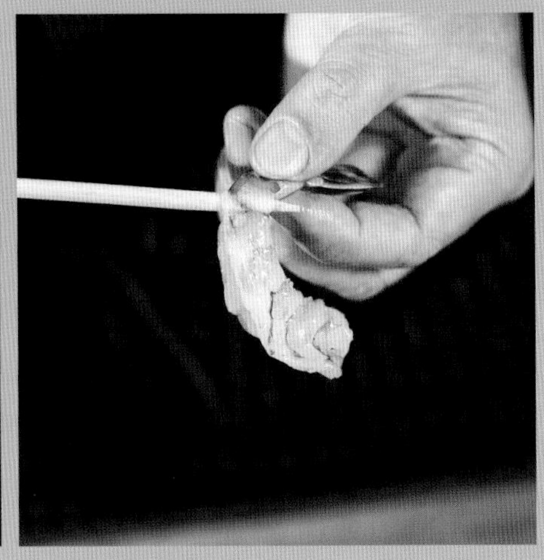

材料

大花蝦……6隻
鮮露筍……200克
冬筍……80克
金華火腿肉……20克

料頭

薑……6片
甘筍……6片
蒜茸……1茶匙

醃料

鹽……1/3茶匙
粟粉……1茶匙
麻油……半茶匙
胡椒粉……1/4茶匙
蛋白……2茶匙

調味料

鹽……1/3茶匙
蠔油……1茶匙
粟粉……半茶匙
清水……1湯匙

做法

1/ 大花蝦切去頭部，去除蝦殼留尾，剔開蝦背少許成蝦球，洗淨，吸乾水分，拌入已混和的醃料待醃味。

2/ 金華火腿肉用清水浸過面蒸約 20 分鐘，盛起，剁成火腿茸。

3/ 冬筍切片，用滾水灼約 5 分鐘去異味；鮮露筍削去表皮，切開兩段。

4/ 在蝦球背部的前端及尾部各穿一小孔，用露筍尖端穿過蝦球的前端及尾部。

5/ 鑊內放入清水 2 杯，下鹽 1 茶匙、砂糖半茶匙、生油 2 茶匙燒熱，放入露筍蝦球灼約 2 分鐘，盛起。

6/ 鑊內下生油 1 茶匙，爆香料頭及冬筍片，再倒入露筍蝦球炒一會，放入調味料用中火快炒，上碟，灑上火腿茸即成。

Chef's Tips

- 大花蝦頭用中油溫炸熟，灑上味椒鹽，可作伴碟之用。味椒鹽在超市有售。
- 露筍不宜選購太粗的品種，約筷子般粗幼為佳。

肉香

好滋味

Meat

濃郁的香料味道，你會愛上嗎？

紫蘇醬油雞

材料

光雞……1 隻
雞蛋……4 個

醬油香料

八角……4 粒
桂皮……20 克
南薑……30 克（切片）
薑肉……60 克（切片）
乾葱頭……6 粒
香茅……2 支（切段）
葱……3 棵
陳皮1 個（浸軟；刮內瓤）

調味料

生抽……500 毫升
印尼甜豉油……200 毫升
清水……10 杯
紹酒……100 毫升
片糖……600 克
鹽……30 克
雞粉……30 克
九層塔……2 棵
乾紫蘇葉……10 片

做法

1/ 大鍋內下生油 2 湯匙，放入醬油香料用中小火炒香，下調味料煮滾，轉慢火煮約 1 小時。

2/ 光雞洗淨，放入紫蘇醬油內煮滾，轉慢火煮約 5 分鐘，關火焗浸約 30 分鐘。

3/ 雞蛋用滾水烚熟，取出去殼，放入醬油內與光雞同浸半小時。

4/ 醬油雞取出斬件上碟，雞蛋切半伴邊，淋上醬油汁享用。

Chef's Tips

- 浸雞後，將醬油汁內的所有材料棄掉，並再次煮滾以免變壞，待涼後可貯存。
- 浸雞時最宜將整隻雞身浸着，可平均受熱，而且味道可滲入雞肉。
- 關鍵在於熄火後浸雞半小時，肉質嫩滑。

在忙碌工作中想找點刺激，黑椒辣味可滿足你嗎？

黑椒雜菌牛柳粒

Chef's Tips

- 如選用較韌的牛部位，醃料可加入少許果酸醃製，如檸檬汁或木瓜汁等，有助軟化肉質。

材料

牛柳……300 克
鮮冬菇……6 朵
蘑菇……10 粒
雞髀菇……100 克
三色甜椒……100 克
黑椒醬汁……2 湯匙
（做法參考 p.15）

料頭

蒜茸……1 茶匙
薑片……10 克

醃料

生抽……2 茶匙
粟粉……2 茶匙
清水……2 湯匙
紹酒……1 茶匙

調味料

蠔油……2 茶匙
砂糖……半茶匙
粟粉……1 茶匙
清水……1 湯匙
黑胡椒粉…1/4 茶匙

做法

1/ 牛柳洗淨，切丁粒，放入醃料拌勻醃約 1 小時。

2/ 所有菇類洗淨，切粒，用滾水稍灼，隔去水分；三色甜椒洗淨，切角。

3/ 鑊燒熱下生油 1 湯匙，將牛柳粒用中慢火煎香至熟，放入料頭用中火炒香，加入黑椒醬汁、三色甜椒及菇類同炒一會。

4/ 調味料拌勻，加入鑊內用大火快炒，即可上碟。

開心脆骨

果仁的脆香，肉汁爆發難以抵擋

材料

一字排骨……400 克
開心果仁……250 克
蒜子……6 粒
薑肉……40 克
低筋麵粉……80 克

醃料

鹽……半茶匙
雞粉……半茶匙
紹酒……2 茶匙
蛋液……2 湯匙
粟粉……3 茶匙
清水……2 湯匙

沙律醬調味料

沙律醬……150 毫升
花生醬……1 湯匙
熱開水……3 湯匙
煉奶……1 湯匙
鮮檸檬汁……1 茶匙

做法

1/ 一字排骨斬約 2 吋長的排骨形，洗淨，瀝乾水分。

2/ 蒜子及薑切碎，與醃料混合，放入排骨拌勻醃 1 小時。

3/ 開心果仁壓碎成幼粒，備用。

4/ 花生醬先用熱開水調開至軟滑，再加入其他調味料拌勻，冷藏備用。

5/ 油鍋內加入生油半鍋，用中火燒熱，排骨表面沾上低筋麵粉，炸至全熟及表面金黃香脆，隔去油分。

6/ 待香脆排骨稍涼，沾上沙律醬，再在表面黏上開心果仁碎，上碟享用。

Chef's Tips

- 香脆排骨剛炸熟盛起時，不要立即沾上沙律醬，否則沙律醬很易溶化。
- 先用熱開水調開花生醬，冷水很難令花生醬溶開來，然後加入其他調味料拌勻，別一次過全部加入。
- 若沙律醬調味料用不完，可放雪櫃一至兩天使用。

仙鶴神針（素翅火腩扣乳鴿）

經典傳統粵菜，用心改良配搭，在家也可品嘗其精髓

材料

乳鴿........................1 隻
仿翅........................100 克
金華火腿肉........................30 克
乾冬菇........................2 朵
火腩......100 克（切小件）
蒜子........................6 粒
薑........................4 片

調味料

鹽........................半茶匙
蠔油........................2 湯匙
老抽........................1 茶匙
砂糖........................半茶匙
柱候醬........................1 湯匙
胡椒粉........................1/4 茶匙
紹酒........................2 茶匙
清水........................2.5 杯

做法

1/ 乳鴿去內臟，洗淨，抹乾水分；仿翅用清水浸軟，隔去水分。

2/ 乾冬菇用清水浸軟、去蒂，用滾水灼約 15 分鐘至腍，切絲，備用。

3/ 金華火腿肉加入清水約 100 毫升，浸過面蒸約 20 分鐘，切幼條，蒸火腿湯留用。

4/ 火腿湯放入鑊內，加入冬菇絲、火腿條及仿翅拌勻煮熱，用粟粉水勾薄芡。

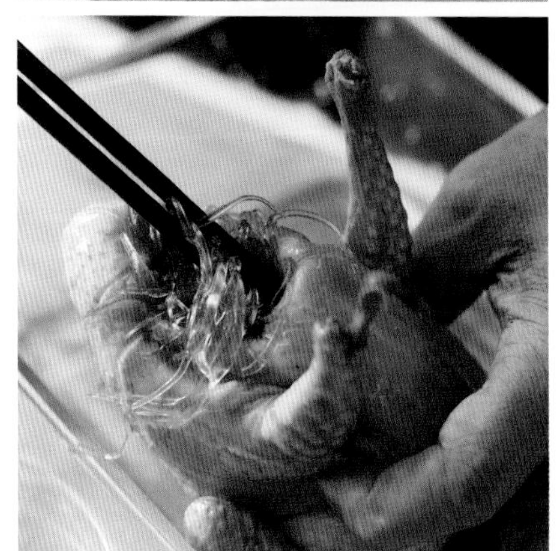

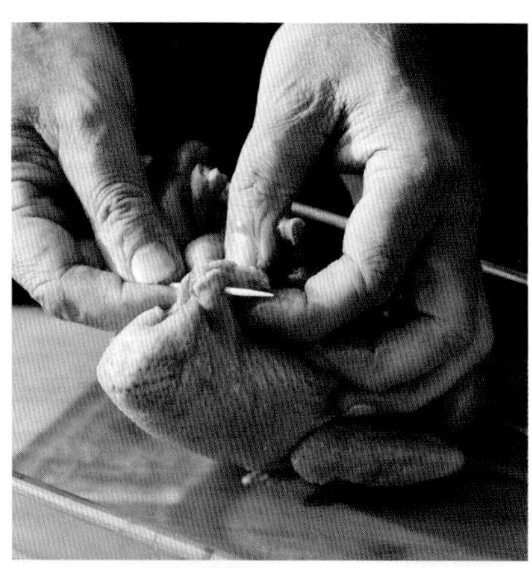

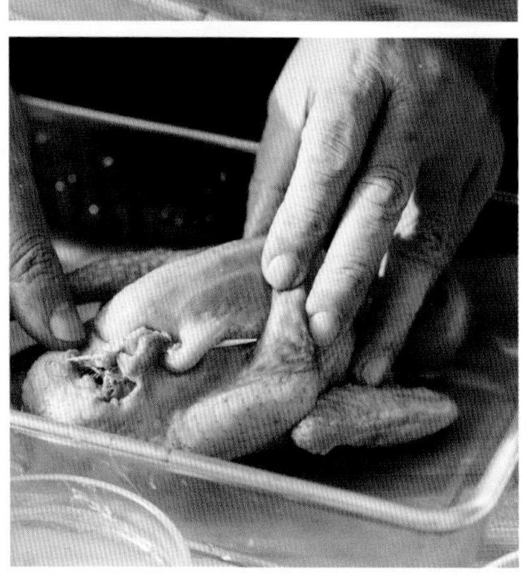

5/ 將以上材料釀入乳鴿腔內填滿，用牙籤封口，用老抽塗抹表皮。鑊燒熱下 1 湯匙生油，用中火將乳鴿表皮稍煎至上色，放入小砂鍋內。

6/ 鑊中下生油 1 湯匙，用慢火煎香蒜子至金黃色，加入薑片及火腩炒香，放入調味料煮熱，倒入砂窩內浸過乳鴿表面。

7/ 砂鍋加上蓋，用小火燜乳鴿約 40 分鐘至腍即可，倒出湯汁用粟粉水勾芡，取出牙籤，剪開乳鴿胸肉，倒上芡汁煲滾享用。

Chef's Tips

・如不是即煮即食，可燜約半小時，關火待涼後冷藏，食用時回鑊蒸約 15 分鐘，用湯汁勾芡即可。

咖喱牛崧煎蛋角

形如金元寶，皮脆肉香，怎能抗拒？

材料

雞蛋……4 個
免治牛肉……160 克
洋葱……80 克（切碎）
蘑菇……8 粒
芫茜……1 棵（切碎）
油咖喱……2 茶匙
牛油……10 克

醃料

鹽……1/4 茶匙
粟粉……1 茶匙
蛋液……2 茶匙
清水……1 茶匙
胡椒粉……適量

調味料

鹽……1/3 茶匙
雞粉……1/3 茶匙
麻油……1/3 茶匙
粟粉……1 茶匙
開水……2 茶匙

Chef's Tips

・炒牛崧及其他材料時，如滲出水分必須瀝乾後才放入蛋液內，以免水分太多難以定型。

做法

1/ 免治牛肉與醃料拌勻，待約半小時。

2/ 蘑菇飛水，切幼粒備用。

3/ 將調味料內的粟粉與開水 2 茶匙調勻，再放入其他調味料拌勻，加入雞蛋 3 個拂打成蛋液。

4/ 鑊燒熱下生油 1 湯匙，用中小火將牛崧炒熟，盛起。

5/ 鑊洗淨及燒熱，下牛油及油咖喱，用中小火炒香洋葱粒、蘑菇粒及牛崧，與芫茜碎一同放入蛋液內拌勻。

6/ 鑊燒熱下生油滑鑊（轉動鑊令油均勻分佈），倒出油分，鑊內餘下約 1 茶匙生油，改小火傾入蛋液料，慢火煎香兩面至熟成圓餅形，切成蛋角。

7/ 取雞蛋 1 個拂打成蛋液，將蛋角蘸蛋液用慢火煎至表面金黃香脆，即可享用。

清甜燉湯，是繁忙生活的調味劑

清燉菜膽獅子頭

Meat

材料

大白菜……8 棵
五花腩肉……500 克
鮮淮山……100 克
金華火腿肉……40 克
上湯……4 杯
（做法參考 p.10）
薑片……40 克

調味料

鹽……2/3 茶匙
雞粉……半茶匙
清水……80 毫升
雞蛋……1 個
紹興酒……1 茶匙
薑米……30 克
胡椒粉……1/4 茶匙
粟粉……2 茶匙

做法

1/ 五花腩肉去皮，洗淨、抹乾水分，切薄片後再切幼條，最後切成幼粒，用刀略剁成肉碎。

2/ 將調味料的水分逐少加入肉料，順一方向攪拌至有黏力，加入其他調味料拌勻，攪至起膠質。

3/ 鮮淮山去皮、洗淨，切幼粒，加入肉餡拌勻，擠成獅子頭肉丸。

4/ 燒滾水後轉小火，放入肉丸稍定型，輕輕盛起放入燉盅。

5/ 金華火腿肉切幼粒；大白菜稍削去大葉，破開兩面，洗淨。火腿粒及大白菜用滾水飛水，盛起放入燉盅內。

6/ 上湯 4 杯放入薑片煮滾，注入燉盅內浸過材料，燉約 1.5 小時即可享用。

Chef's Tips

- 五花腩肉宜選購 3 分肥 7 分瘦、無筋膜的為最佳。
- 腩肉去皮、洗淨後，最好放入雪櫃冷藏至略硬身，較容易切成薄片及幼粒。
- 每粒肉丸要擠得大小均勻，每個約 60 至 80 克為佳。
- 如不喜愛鮮淮山，可用馬蹄肉代替。

椒醬肉

粒粒爽口，是伴麵的理想配搭

材料

五花腩肉……300 克
沙葛……120 克
菜脯……100 克
青、紅甜椒100克（切粒）
五香豆乾……3 件
蝦乾……60 克
洋葱粒……50 克
炸花生……50 克

料頭

蒜茸、薑片、葱粒各適量

調味料 1（燜腩肉）

鹽……半茶匙
雞粉……半茶匙
蠔油……2 湯匙
老抽……2 茶匙
冰糖……15 克
八角……3 粒
薑……3 大片
葱……2 棵
清水……3 杯

調味料 2

海鮮醬……2 茶匙
豆瓣醬……3 茶匙
柱侯醬……1 茶匙
蠔油……1 湯匙
鹽……1/3 茶匙
砂糖……1 茶匙
鎮江香醋……2 茶匙
老抽……1 茶匙
清水……2 湯匙
粟粉……1 茶匙

做法

1/ 五花腩肉飛水，洗淨。煮滾調味料 1，放入腩肉用慢火燜約 45 分鐘，切粒備用。

2/ 蝦乾用清水浸軟，與其他材料切粗粒。

3/ 沙葛、五香豆乾及菜脯粒用熱水浸熱，隔水。

4/ 燒熱鑊後下油 2 茶匙，爆香料頭，再加入沙葛、菜脯、甜椒粒、豆乾、蝦乾、洋葱粒炒香。

5/ 加入腩肉粒及已拌勻的調味料 2 用大火快炒均勻，最後灑上炸花生，上碟食用。

Chef's Tips

- 如不選用五花腩肉，可用半肥瘦叉燒代替，省卻燜腩肉的步驟。

荷香荔芋扣肉

愛芋頭綿香，夾雜肉脂及濃醬，入口香腴

材料

五花腩……600 克
芋頭……500 克
蒜茸……1 湯匙
乾荷葉……1 塊

調味料

清水……200 毫升
南乳……2 大磚
柱侯醬……1 湯匙
鹽……1/3 茶匙
砂糖……3 湯匙
蠔油……3 湯匙
紹酒……2 湯匙

做法

1/ 五花腩洗淨，用滾水焓約 30 分鐘，盛起洗淨，表面用老抽塗抹，煎香表皮上色。

2/ 芋頭去皮，切成長方形厚片（約 3 x 5厘米），用中油溫炸香備用。

3/ 乾荷葉用清水浸軟，用滾水稍灼去異味，抹乾水分，鋪在大窩碗內。

4/ 腩肉切成與芋頭大小的厚片。

5/ 在鑊內下生油 1 湯匙，加入蒜茸及調味料爆香，盛起。

6/ 將腩肉及芋頭片相間排入碗內，最後倒入調味汁用荷葉包好，用中火隔水蒸約 2.5 小時即可。

Chef's Tips

- 芋頭要切成厚片，若太薄當蒸製時很易散爛。
- 芋頭厚片可不油炸，改用油煎至兩面稍硬身。
- 細粒的腩肉碎可排入芋頭片與腩肉的空隙位置，令蒸出來的扣肉更緊密。
- 如扣肉做好後不立即吃，可放雪櫃貯存一天，享用前再蒸熱進食，更加入味。

栗香、煙肉焦脆，你會驚嘆如此滋味！

金栗焗雞中翼

材料

- 雞中翼......10 隻
- 大栗子......10 粒
- 煙肉片......10 片
- 蜜糖......4 湯匙

醃料

- 海鮮醬......3 湯匙
- 柱候醬......1 湯匙
- 蠔油......2 湯匙
- 鹽......1/3 茶匙
- 雞粉......1 茶匙
- 砂糖......4 湯匙
- 麻油......1 湯匙
- 紹酒......1 湯匙
- 蒜茸......2 湯匙
- 薑片......20 克
- 雞蛋......1 個

Chef's Tips

- 如雞中翼不是太大，可減少烘焗時間，避免過火焦燶。

做法

1/ 栗子放入滾水內煲熟，去殼除皮，備用。

2/ 雞中翼去掉翼骨，醃料混合後放入雞中翼醃約半小時。

3/ 煙肉用溫水浸泡一會去鹹味，抹乾，備用。

4/ 將栗子釀入雞中翼內，用煙肉片包捲雞中翼，用牙籤穿好固定。

5/ 預熱焗爐，雞中翼放在焗盤，用 220℃烘約 5 分鐘，轉 150℃烘約 8 分鐘，取出。

6/ 在雞中翼表面塗上蜜糖，再用 200℃烘約 2 分鐘至焦香，取出上碟享用。

自家製的馬拉盞，多一份味道，多一份心思

馬拉盞椒絲蒸腩肉片

材料

五花腩肉......300 克
紅尖椒......1 隻
薑肉......30 克
蝦乾......80 克
葱......1 棵

調味料

馬拉盞......1 湯匙
（做法參考 p.14）
鹽......1/4 茶匙
砂糖......2 茶匙
粟粉......1 茶匙
麻油......半茶匙

做法

1/ 五花腩肉洗淨，切薄片備用。

2/ 紅尖椒、薑肉及葱切絲備用。

3/ 蝦乾用清水浸軟，隔水後放入腩肉片內，加入薑絲、紅椒絲及調味料拌匀，鋪於碟上。

4/ 放於鑊內用中火蒸約 12 分鐘，取出灑上葱絲享用。

Chef's Tips

- 五花腩肉連皮切成薄片，蒸熟後口感更佳。
- 如要較易切成腩肉薄片，洗淨後抹去水分及冷藏，稍硬身時才容易切片。

惹味的冷盤，是飯前的開胃菜

榨菜滷牛腱

材料

牛腱......500 克
榨菜......2 個
紅尖椒......1 隻

榨菜調味料

砂糖......2 湯匙
麻油......1 湯匙
豆瓣醬......2 茶匙

滷水香料

甘草......10 克
紅穀米......10 克
草果......3 粒
沙薑......6 克
花椒......5 克
八角......4 粒
陳皮......1/3 片
蒜子......6 粒
薑片......40 克
葱......2 棵
芫茜......3 棵

調味料

清水......5 杯
冰糖......100 克
生抽......60 毫升
鹽......半茶匙
雞粉......2 茶匙
五香粉......1 茶匙
紹酒......40 毫升

做法

1/ 榨菜洗淨，切丁粒，用溫水浸泡去鹹味，隔水，與榨菜調味料拌勻備用。

2/ 大鍋內下生油 1 湯匙，放入滷水香料慢火炒香，加入調味料煮滾，轉慢火煲約 1 小時。

3/ 牛腱用滾水焓約 5 分鐘，取出洗淨，放入滷水內煮滾，轉慢火煲約 1 小時，關火浸焗至涼，盛起。

4/ 紅尖椒切幼粒，牛腱切丁粒，與榨菜粒拌勻，淋上少許滷水汁，上碟享用。

Chef's Tips

- 如原條牛腱太大，可先切開兩件才飛水，煲焗時易於入味。

豆

飄 香

粒粒金黃的滋味

七味黃金磚

材料

板豆腐……1 件
鹹蛋黃……4 個
牛油……80 克

鹽水料

溫水……1 杯
幼鹽……1 茶匙

七味脆炸漿

低筋麵粉……200 克
泡打粉……20 克
日本七味粉……2 茶匙
鹽……1/3 茶匙
清水……約 200 毫升
生油……1 湯匙（後下）

Chef's Tips

- 豆腐放入鹽水略浸，目的是令豆腐的蛋白質結實，水分不易釋出。
- 脆炸漿拌勻後不宜立即使用，需靜候半小時或以上，使漿粉發酵才達到鬆脆的效果。
- 炒鹹蛋黃前，先將牛油用慢火煮溶，才加入鹹蛋黃炒勻，否則難以呈現流沙的狀況。

做法

1/ 脆炸漿材料拌勻，靜待約半小時，加入生油 1 湯匙攪勻，備用。

2/ 鹹蛋黃隔水蒸約 15 分鐘至熟，壓碎備用。

3/ 板豆腐切開約 1 吋立方體，放入鹽水略浸 20 分鐘，隔水，用乾布吸乾水分。

4/ 鑊內下油燒熱，轉中火，豆腐黏上脆炸漿，炸至金黃及脆身即可盛起。

5/ 鑊內放入牛油用慢火煮溶，加入碎蛋黃炒勻，再放入已炸豆腐粒拌勻，上碟。

嫩滑的蛋白豆漿遇上爽蝦，宴客之選

白雪尋龍

材料

蝦仁……300 克
雞蛋……8 個
淡豆漿……150 毫升
粟米片……120 克
金華火腿茸……10 克
（做法參考 p.9）

蛋白調味料

鹽……1/3 茶匙
雞粉……1/3 茶匙
粟粉……2 茶匙
清水……3 茶匙

醃料

鹽……半茶匙
粟粉……2 茶匙
麻油……1 茶匙
胡椒粉……適量

Chef's Tips

- 以淡豆漿代替慣常使用的鮮奶，令不適合飲用鮮奶者也能品嘗，而且豆漿的水分少、澱粉質多，容易炒至凝固。
- 將蛋白淡豆漿炒至滑嫩的技巧：毋須使用太大火，只要炒鑊的熱度足夠，最後階段可熄火翻炒。
- 餘下的蛋黃加入清水 3 湯匙拌勻，用慢火燒熱平底不黏鑊，倒入蛋黃液煎成薄蛋皮，切絲後伴邊裝飾。

做法

1/ 雞蛋 8 個分開蛋白及蛋黃；用蛋白調味料的清水調溶粟粉，加入其他調味料拌勻，放入蛋白內攪勻。

2/ 蝦仁用粟粉洗淨，用乾布吸乾水分，加入醃料拌勻。

3/ 淡豆漿倒入已調味的蛋白拌勻。

4/ 蝦仁飛水，用中油溫泡油，盛起。

5/ 將淡豆漿混合物倒入鑊內，用鑊鏟以慢火炒至 8 成熟，放入蝦仁同炒至全熟。

6/ 碟內鋪上部分粟米片，放入炒熟蛋白豆漿，灑上火腿茸，粟米片拌邊即成。

熱辣辣上桌，還不快快舉筷

山根蝦籽豆腐煲

材料

山根……10 個
乾冬菇……6 朵
布包豆腐……2 件
小棠菜……8 棵
蝦籽……2 茶匙

調味料

鹽……1/3 茶匙
蠔油……1 湯匙
老抽……1 茶匙
清水……半杯

做法

1/ 山根用滾水灼軟，盛起，漂冷水，沖去油分。

2/ 冬菇浸軟，洗淨，切片備用。

3/ 布包豆腐每件切開 6 小件，放入高油溫（約 170℃）炸至金黃，盛起。

4/ 鑊中下生油 1 茶匙炒香蒜茸半茶匙、薑數片，放入調味料煮熱，加入所有材料用慢火燜約 2 分鐘，拌入 2 湯匙粟粉水勾芡，放入瓦煲內。

5/ 小棠菜伴邊，用中火煲滾，最後灑上蝦籽即可品嘗。

Chef's Tips

- 蝦籽可選購樽裝產品，若選用散裝蝦籽，宜先用乾鑊小火炒香蝦籽，待涼後貯存於小樽，留待日後使用。
- 豆腐也可用中火煎香各面，但燜煮時緊記別攪拌太多，以免弄散豆腐。
- 炸後的布包豆腐較嫩滑；硬豆腐的口感則結實。

腐皮鯪魚卷

鬆脆、爽彈、滋味，齒頰留香

材料

腐皮……1 塊
鯪魚膠……300 克
臘腸……1 條
膶腸……1 條
蝦米……40 克
葱花……10 克
芫茜碎……10 克

脆炸漿

低筋麵粉……200 克
泡打粉……20 克
鹽……1/3 茶匙
清水……約 220 毫升
生油……1 湯匙（後下）

做法

1/ 脆炸漿混合均勻，靜待半小時，備用。

2/ 蝦米浸水至軟，剁碎後放入鯪魚膠攪勻，加入葱花及芫茜碎搓勻，略撻數次。

3/ 臘腸及膶腸飛水約 1 分鐘，抹去水分，臘腸及膶腸切開兩邊；腐皮剪成 2 塊長方形（寬度與臘腸長度一樣）。

4/ 鯪魚膠平均鋪在腐皮上，排上臘腸及膶腸，捲起腐皮緊包鯪魚膠表面，再用錫紙包好，蒸約 8 分鐘至熟，取出待涼。

5/ 鑊內注入生油用中火燒熱，鯪魚卷蘸上脆炸漿，放入熱油炸至金黃香脆，盛起，吸去油分，切件上碟。

Chef's Tips

- 可在市場購買已調味的鯪魚膠，較為方便，可自行再加添其他配料。
- 蒸鯪魚卷時建議用錫紙包卷或用手布蓋住魚卷，以免倒汗水滴在魚卷上，弄濕腐皮表面。

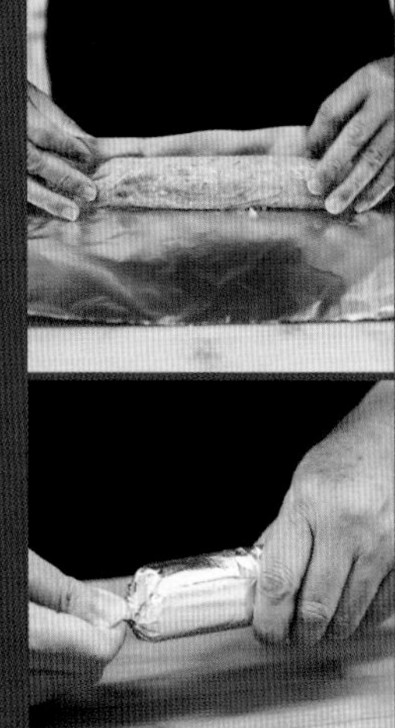

一絲絲，入口綿滑，讚嘆精湛的刀功

千絲萬縷

材料

布包豆腐……1件
鮮腐皮……1片
冬瓜……200克
萵筍……200克
甘筍……120克
金華火腿茸……適量
上湯……2杯
蛋白……1個

調味料

鹽……半茶匙
胡椒粉……1/4茶匙
麻油……1/3茶匙

粟粉芡料

粟粉……2湯匙
清水……4湯匙

做法

1/ 布包豆腐切薄片，再切成絲，用溫開水浸泡。

2/ 其他材料切薄片，再切幼絲，用滾水加鹽 1 茶匙灼熟，盛起。

3/ 鑊中倒入上湯煮熱，加入調味料拌勻，下材料煮熱（豆腐絲除外），用粟粉料勾芡。

4/ 湯羹煮至濃稠後，豆腐絲瀝乾水分，放入湯內輕拌，加入蛋白攪拌煮成蛋花即熄火，灑上火腿茸享用。

Chef's Tips

- 豆腐切絲沒有想像中困難，只要刀鋒利，並選用大面積的菜刀，容易切成幼絲。

來一口微暖的鹹豆花，配搭清新

黃金豆花

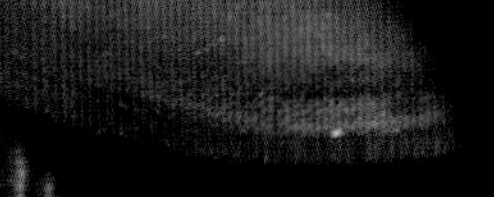

材料

淡豆腐花..........................2 碗
鹹蛋黃...............................4 個
南瓜.............................300 克
金華火腿肉..................30 克
葱......................1 棵（切粒）
清水..................................半杯

調味料

鹽................................1/3 茶匙
麻油...........................1/3 茶匙

做法

1/ 金華火腿洗淨，用半杯清水浸過面，蒸約半小時，隔水盛起，火腿汁留用，火腿肉剁成茸。

2/ 鹹蛋黃蒸約 8 分鐘，壓碎；南瓜切粒，蒸約 5 分鐘，備用。

3/ 鑊中倒入火腿汁、南瓜粒及鹹蛋黃，加入調味料煮熱，下 2 湯匙粟粉水勾芡。

4/ 淡豆腐花盛於碗內，舀上南瓜鹹蛋黃，灑上火腿茸及葱花，趁暖享用。

Chef's Tips

如要將淡豆花翻熱，可用慢火蒸約 5 分鐘，但別用猛火蒸，會令豆花過熟不滑。

甘苦的韻味，再三細味品嘗

欖菜涼瓜炒豆乾

材料

豬肉碎......150 克
五香豆乾......3 件
涼瓜......300 克
欖菜......2 湯匙

料頭

蒜茸......半茶匙
薑粒......半茶匙
葱粒......半茶匙
豆豉......1 茶匙

醃料

鹽......1/4 茶匙
粟粉......2 茶匙
清水......4 茶匙

芡汁料

鹽......1/3 茶匙
蠔油......2 茶匙
老抽......1 茶匙
砂糖......1 茶匙
清水......3 茶匙
粟粉......1 茶匙

做法

1/ 醃料混合，放入豬肉碎中拌勻。

2/ 涼瓜開邊、去籽，放入滾水灼約 3 分鐘，盛起漂冷水，吸乾水分，切件。

3/ 五香豆腐切丁粒，用滾水浸熱，隔水備用。

4/ 鑊燒熱後下油 1 湯匙，下豬肉碎炒熟盛起。

5/ 鑊內再下生油 1 茶匙爆香料頭及欖菜，放入其他材料炒香，最後加入芡汁料炒勻即可。

Chef's Tips

- 如想涼瓜味道不太甘苦，飛水後用茶匙刮去白瓤，洗淨即可。
- 炒吃的可挑選長型涼瓜；燜煮則選用短身的雷公鑿。

靈活巧妙的配搭，成為你獨一無二的作品

好市琵琶豆腐

材料（可製成 16 件）

布包豆腐……1 件
＊蝦膠……180 克
乾蠔豉……10 隻
冬菇……4 朵
雞蛋……1 個
芫茜……1 棵
金華火腿茸……20 克

調味料 1（燜蠔豉及冬菇）

薑……4 片
葱……2 棵
鹽……半茶匙
蠔油……1 茶匙
砂糖……1 茶匙
紹酒……1 湯匙
清水……2 杯

調味料 2

鹽……半茶匙
蠔油……1 湯匙
雞蛋黃……1 個
粟粉……3 湯匙
胡椒粉……1/4 茶匙
麻油……半茶匙

做法

1/ 冬菇及蠔豉用清水浸軟，冬菇去蒂洗淨，蠔豉除沙洗淨。

2/ 煮熱調味料 1，放入冬菇及蠔豉用慢火燜約半小時，隔去水分，切幼粒。

3/ 芫茜洗淨，摘葉備用，芫茜梗切碎。

4/ 豆腐壓去水分，放入碗內加入調味料 2 拌勻，下冬菇及蠔豉粒拌勻，拌入蝦膠撻勻，即成豆腐蝦膠。

5/ 瓦湯匙塗上生油，舀入豆腐蝦膠至平口，鋪上芫茜葉及火腿茸，用中火蒸約 8 分鐘，取出待涼，用小刀剔出豆腐蝦膠成為琵琶形狀。

6/ 琵琶豆腐蘸上蛋液，用中火煎香兩面，上碟享用。

預備蝦膠

材料

- 蝦肉............................180 克
- 粟粉............................2 茶匙
- 鹽............................1/3 茶匙
- 麻油............................1/3 茶匙

做法

1/ 蝦肉用粟粉 1 湯匙洗擦，漂水洗淨，用乾布吸乾水分。

2/ 用刀拍散蝦肉，再用刀背剁爛至蝦茸。

3/ 蝦茸放入大鍋內順方向攪拌至有黏力，再放入鹽及麻油攪勻，攪撻至有膠質即可。

Chef's Tips

- 琵琶豆腐的調味料包含蛋黃及粟粉，有令豆腐蝦膠結實的效果。
- 蒸後的豆腐蝦膠會漲起來，要稍涼一會令其下降，容易移離瓦匙。
- 布包豆腐未使用時，浸泡清水內並放入雪櫃，以免發出異味。
- 可運用自己的創意，隨意加入任何材料如煙肉，令美食變得不平凡。

咬入口，爽脆！是炎夏的透心涼菜

香芹枝竹

材料

材料	份量
西芹	250 克
枝竹	2 條
白背木耳	2 朵
紅尖椒	1 隻

涼拌調味料

調味料	份量
鹽	半茶匙
蠔油	2 湯匙
豆瓣醬	2 茶匙
砂糖	1 茶匙
麻油	2 茶匙
蒜茸	1 茶匙

做法

1/ 枝竹及木耳用溫水浸軟，枝竹剪成小段；木耳切絲。

2/ 西芹刨去表面硬皮，洗淨，切小段，再切成幼條。

3/ 煮滾清水 2 杯，加入鹽 2 茶匙、生油 1 茶匙，放入西芹、枝竹、木耳灼熟，盛起，用冷開水浸涼，瀝乾水分，冷藏 1 小時。

4/ 涼拌調味料混和，倒入材料內拌勻，上碟食用。

Chef's Tips

・涼拌醬汁享用時才拌入，否則太早與材料調勻，因含鹽分，冷藏後容易釋出水分，令味道變淡。

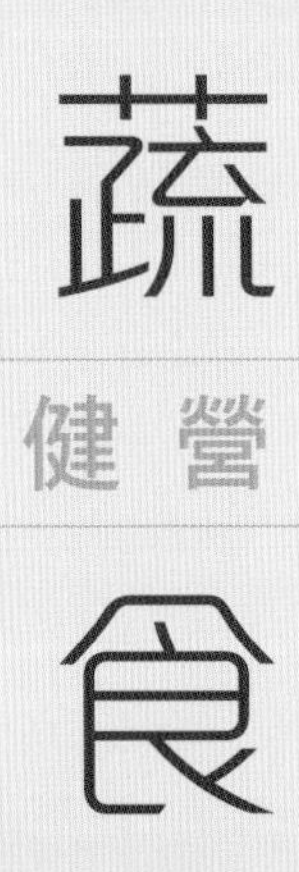
蔬
健營
食

用心做菜，吃到的是誠意與心思

釀蓮藕燜腩肉

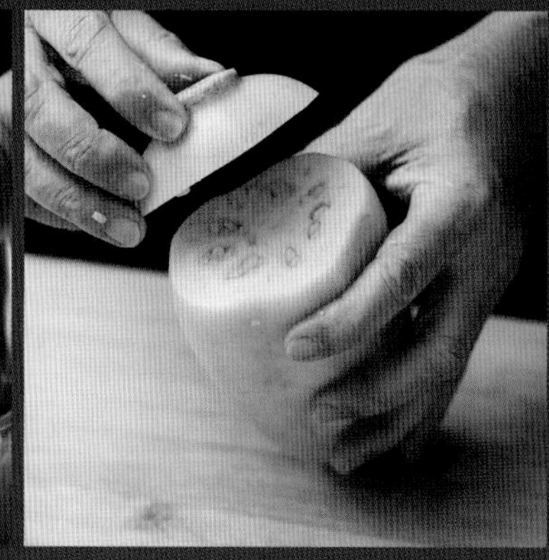

材料

五花腩......600 克
蓮藕......400 克
去殼綠豆......150 克

料頭

乾葱......10 粒
大薑片......4 片
葱......4 棵（切段）
八角......6 粒
陳皮......1/3 片

調味料

生抽......1 杯
花雕酒......1 杯
蠔油......2 湯匙
老抽......2 湯匙
冰糖......120 克
鹽......1 茶匙
雞粉......1 茶匙
柱候醬......2 湯匙
清水......12 杯

做法

1/ 五花腩用熱水煮約 20 分鐘，取出，漂冷水洗淨。

2/ 去殼綠豆用清水浸約 1 小時，放入熱水煲約 4 分鐘，隔水備用。

3/ 蓮藕去皮，洗淨，在尖端部位切開，將綠豆釀入蓮藕的小孔內，將尖端部分蓋上，用竹籤固定。

4/ 煲內下少許生油爆香料頭，放入調味料煮熱，加入五花腩肉及蓮藕用慢火燜約 1.5 小時，取出腩肉，待涼後冷藏。蓮藕再燜 30 分鐘至腍即可。

5/ 腩肉及蓮藕切成厚片，排在窩碟內，淋上腩肉汁並蒸約 15 分鐘即成。

Chef's Tips

- 綠豆釀入蓮藕時，用幼筷子輔助可壓緊綠豆，但釀入時切勿太多及太壓迫，以免綠豆熟透膨漲時擠至蓮藕爆裂。
- 釀綠豆時要花點耐性，用筷子壓入並慢慢轉動，令綠豆落入蓮藕內，騰出空位後可再釀入綠豆。
- 腩肉及蓮藕切厚片時如發現不夠腍身，可增加蒸熱的時間，蒸至腍身即可。

自家醃製的鹹香五花腩，濃香、滋味！

鮮淮山炒鹹肉

材料

五花腩肉......300 克
鮮淮山......200 克
香芹......100 克
鮮冬菇......4 朵

醃鹹肉料

鹽......2 湯匙
砂糖......1 茶匙
紹酒......2 湯匙
五香粉......1 茶匙
胡椒粉......1/3 茶匙

料頭

蒜茸......半茶匙
薑......6 片

調味料

鹽......1/3 茶匙
蠔油......1 茶匙
粟粉......1 茶匙
清水......2 茶匙

做法

1/ 五花腩切成長塊，洗淨，醃鹹肉料混合，放入五花腩肉拌匀，用保鮮袋包好冷藏一天。

2/ 取出五花腩，洗去鹽分，用滾水焓約 15 分鐘，盛起，切薄片。

3/ 鮮淮山去皮，洗淨及切片；香芹摘去葉片，香芹梗切小段；鮮冬菇切片。

4/ 鮮淮山、香芹、鮮冬菇放於鑊內，下生油 2 茶匙、鹽半茶匙、清水 3 湯匙，用中火炒熟，盛起。

5/ 下生油 2 茶匙，放入腩肉片炒熟，加入料頭用中火爆香，下其他配料同炒，調味料混和後，灑入鑊內用中火炒匀，上碟享用。

- 如不喜愛五花腩肉的豬皮，可先去皮才切成薄片。
- 鮮淮山去皮時建議帶上手套，否則容易令手部痕癢。

中式的香料搭配日本調味料，口味耳目一新

清酒黃瓜鮮鮑片

Chef's Tips

- 選用小青瓜較合適，因為粗大的青瓜較難入味。
- 清酒不宜煲煮加熱，令酒味揮發散去。

材料

鮮鮑魚……6 隻
小青瓜……2 條

浸漬調味料

八角……2 粒
杞子……20 克
清水……1 杯
鹽……半茶匙
砂糖……半茶匙
日本木魚片……20 克
日本清酒100 毫升（後下）

做法

1/ 浸漬調味料用慢火煮熱，待涼，加入清酒拌勻備用。
2/ 小青瓜直切成兩邊，切小段，用鹽 2 茶匙拌勻醃約 3 分鐘，用清水洗去鹽分。
3/ 鮮鮑魚洗擦外殼污漬，蒸約 6 分鐘至熟，浸泡冰水待涼，除殼及內臟，洗淨備用。
4/ 將小青瓜及鮑魚肉浸漬在調味料，冷藏一天，翌日取出切片享用。

剔透亮白的外表，啖啖清甜，肉香滿溢

晶瑩小丸子

材料

豬肉丸漿..................300 克
（做法參考 p.12）
白蘿蔔.......................500 克
荷葉...............................1 塊
金華火腿茸.................20 克
葱花.............................適量

做法

1/ 將豬肉丸漿唧成 12 粒小丸。

2/ 白蘿蔔去皮、切薄片，再切成幼絲，用 2 茶匙幼鹽拌勻，醃約 3 分鐘軟身，漂水洗去鹽分。

3/ 荷葉飛水，抹乾水分，鋪碟上。

4/ 蘿蔔絲瀝乾水分，豬肉丸黏滿蘿蔔絲輕輕捏緊，放於荷葉上蒸約 10 分鐘，取出，灑上火腿茸及葱花即成。

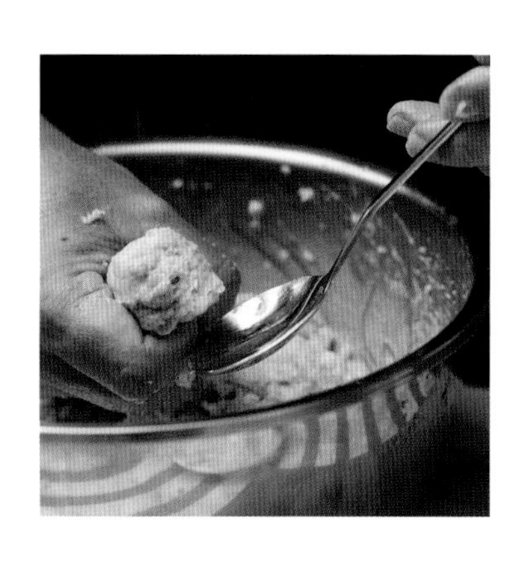

Chef's Tips

- 如買不到鮮荷葉，可用乾荷葉代替，但要先將乾荷葉用清水浸軟才飛水使用。
- 如想菜餚加點色彩，可加入紅蘿蔔絲混合，做法與白蘿蔔絲相同，用幼鹽略醃容易蒸至軟身。

清爽的素食，給生活一點清新感覺

豆苗竹笙卷

材料

乾竹笙............................10 棵
豆苗..............................400 克
金菇..............................200 克

調味料

上湯半杯（做法參考 p.10）
鹽..............................1/3 茶匙
粟粉..............................2 茶匙
胡椒粉..............................適量

Chef's Tips

- 浸發乾竹笙時最宜用溫淡鹽水，回軟後用清水反覆漂洗，去除幼砂，至表面色澤潔白。
- 野生竹笙色澤較微黃，氣味清香，價格較高。人工培植的竹笙色澤較潔白，氣味帶少許琉璜味，需用滾水稍灼去其氣味，價格便宜。
- 剪出來的竹笙網狀部分可保留煮成湯羹。
- 除選用豆苗外，葉型的蔬菜如菠菜也可釀進竹笙卷內。

做法

1/ 竹笙用淡鹽水浸軟約 2 小時，剪去菌柄頂部及網狀部分，保留菌柄烹調。

2/ 竹笙放入滾水稍灼去除異味，瀝乾水分，用上湯 1 杯煨味，備用。

3/ 金菇切去尾段根部，再切小段搓散，備用。

4/ 豆苗洗淨、切碎。鑊內下生油 1 湯匙，放入豆苗及鹽 1 茶匙，用中火炒熟，壓去水分。

5/ 竹笙隔去上湯，釀入豆苗呈飽滿排放碟上，隔水蒸約 6 分鐘。

6/ 金菇碎用中油溫炸脆（約 150℃），隔油盛起。

7/ 取出竹笙卷，倒去水分，煮熟調味料成芡汁，扒在竹笙卷上，灑上炸金菇即成。

鬆化的芋茸、鮮甜的蟹肉，完美絕配！

珊瑚荔茸盞

材料

＊荔茸 900 克
（材料參見＊）
蟹肉 100 克
雞蛋 1 個（拂勻）

調味料

鹽 1 茶匙
砂糖 2 茶匙
雞粉 1 茶匙
五香粉 半茶匙
麻油 1 茶匙

蟹肉芡調味料

鹽 1/3 茶匙
上湯半杯（做法參考 p.10）
粟粉 3 茶匙
胡椒粉 1/5 茶匙

＊荔茸材料

芋頭 600 克
澄麵 100 克
固體菜油 80 克

Vegetable

做法

1/ 芋頭去皮，切小件，蒸約 20 分鐘至腍，稍涼後壓散成茸。

2/ 澄麵加入滾水 100 毫升拌勻，搓成半熟麵糰，加入芋茸、固體菜油及調味料搓勻成芋茸。

3/ 芋茸分成約 10 小份麵糰，搓成棋子形，用中油溫（約 150℃）炸至起酥，轉大火（約 180℃）炸脆至金黃色成荔茸盞。

4/ 上湯及蟹肉煮滾，調味料混合後加入煮熟成蟹肉芡，慢慢加入蛋液煮熟成芡汁，將芡汁淋在荔茸盞中間，上碟享用。

Chef's Tips

- 芋頭蒸熟後搓芋茸時，如發現有芋粒不散應拿出棄掉，以免影響質感。
- 炸芋茸盞時如油溫太低，會令芋茸炸至鬆散不成形，故需要改變油溫炸出鬆化的口感。

粒粒黃金蛋，滲出瑤柱香

銀芽瑤柱桂花翅

材料

仿魚翅……150克
芽菜……200克
乾瑤柱……5粒
雞蛋……4個
芫茜……1棵

調味料

鹽……1/3茶匙
胡椒粉……1/4茶匙
麻油……半茶匙

Chef's Tips

- 初次炒芽菜時不宜加水，而且不宜炒至全熟，否則加入蛋液同炒時會滲出水分，口感也不爽脆。
- 這菜式的特色是將蛋液炒至乾身及呈蛋碎粒狀，外觀似桂花，而且蛋香四溢。

做法

1/ 乾瑤柱用清水浸過面約 2 小時，連水蒸 45 分鐘，拆成絲，瑤柱水留用。

2/ 仿魚翅用滾水稍灼，取出放於瑤柱水浸泡至入味。

3/ 鑊燒熱放入生油 1 湯匙，芽菜落鑊用中火快炒，下鹽 1 茶匙炒勻至芽菜 6 成熟，隔去水分。

4/ 雞蛋與調味料拂勻；燒熱鑊並加入生油 1 湯匙，放入蛋液用中慢火炒至半熟，加入仿魚翅及瑤柱絲，轉中火用鑊鏟炒至乾身及香氣溢出。

5/ 最後放入芽菜快炒均勻即可上碟，灑上芫茜即可。

簡單的食材，吃出不平凡之味

味噌蘿蔔環

材料

白蘿蔔環......10 個
鹹蛋黃......2 個
葱花......適量

燜蘿蔔調味料

味噌......2 湯匙
木魚片......20 克
八角......3 粒
老抽......2 茶匙
薑......4 片
鹽......半茶匙
冰糖粒......20 克
清水......3 杯

粟粉芡

粟粉......1 湯匙
清水......2 湯匙

Vegetable

做法

1/ 白蘿蔔去皮，切成圓形棋子狀，洗淨。
2/ 鹹蛋黃蒸熟，放於小鋼篩用湯匙壓散成幼粒。
3/ 煮滾調味料，放入蘿蔔環用慢火燜約 40 分鐘。
4/ 蘿蔔環放於碟上，湯水隔渣，取一杯湯與粟粉水勾芡，扒在蘿蔔環上，最後灑上鹹蛋幼粒及葱花即成。

Chef's Tips

- 這道菜不宜選用太粗的蘿蔔，體形較幼而直身的蘿蔔較合適。
- 如要蘿蔔環形狀統一，建議用鋼圓模壓出棋子形狀的蘿蔔環。

濃濃的蒜香，一吃已叫人難忘

蒜香粉絲蒸勝瓜

材料

勝瓜......1條（約400克）
粉絲......1小包
蒜子......80克
薑......2大片

調味料

生油......2湯匙
麻油......1茶匙
鹽......1/3茶匙
砂糖......1/3茶匙
蠔油......1湯匙
生抽......1茶匙

做法

1/ 將清水2杯放入鍋內，下鹽半茶匙及蠔油1茶匙煮熱，稍涼，放入粉絲浸軟。

2/ 勝瓜相間地削去表皮，直切開邊成4條，切去瓜瓤、切段，洗淨，瀝乾水分。

3/ 蒜子剁碎成蒜茸；薑片切碎成小粒。

4/ 鑊燒熱放入調味料的生油，用慢火炒香薑粒及蒜茸，加入其他調味料混合。

5/ 粉絲隔去水分，放於碟上，將蒜茸調味料與勝瓜拌勻，排放粉絲上，蒸約8分鐘，即可享用。

Chef's Tips

- 浸發粉絲不宜用太熱的湯水，令粉絲過度漲發，烹煮時容易過腍。
- 勝瓜表皮稍硬，不宜保留太多，而且草青味較濃。

米

綿香

粒

Cereals

綿軟的瓦煲飯，重拾傳統的手藝與滋味

蝦乾排骨煲仔飯

材料

腩排......250 克
蝦乾......60 克
冬菇......6 朵
紅棗......6 粒
白米......300 克

醃料

鹽......1/3 茶匙
蠔油......2 茶匙
粟粉......2 茶匙
清水......4 茶匙
薑......6 片
生油......2 茶匙

豉油汁

生抽......3 湯匙
老抽......2 茶匙
砂糖......2 茶匙
生油......1 茶匙
麻油......1 茶匙
開水......3 湯匙
胡椒粉......適量
芫茜......1 棵

Cereals

Chef's Tips

- 放入排骨料時要留意米粒的水分，不要太乾時加入排骨料，否則難以熟透。
- 轉慢火焗時最宜反覆移動砂鍋，以免爐具的中間火位集中在同一位置，容易燒焦。
- 如飯粒太乾而未熟透，可在米飯灑下開水，用慢火焗透至飯粒完全成熟。

做法

1/ 蝦乾洗淨，用清水浸軟。

2/ 冬菇浸軟，去硬蒂，切片；紅棗洗淨，去核備用。

3/ 腩排斬成小件，洗淨，加入醃料、冬菇、紅棗及蝦乾拌勻，醃約半小時。

4/ 白米洗淨，隔水，放入砂鍋內，加入清水 300 毫升，用中慢火煮至米粒水分收少，立即加入排骨料鋪平在飯面，加蓋轉慢火煲約 12 分鐘，至飯粒完全熟透，關火焗約 6 分鐘。

5/ 豉油汁混合煮滾，取出芫茜棄掉，吃時加入豉油汁。

外脆內軟，鹹香誘人

香煎糯米卷

材料

糯米............................300 克
臘腸.................................1 條
膶腸.................................1 條
臘肉...............................半條
冬菇.................................6 朵
蝦米...............................50 克
芫茜................1 棵（切碎）
葱.....................1 棵（切粒）
雞蛋液..............................2 個
紫菜片..............................4 片

調味料

生抽..............................1 湯匙
蠔油..............................1 湯匙
老抽..............................1 茶匙
雞粉............................半茶匙
開水..............................1 湯匙
麻油..............................1 茶匙
胡椒粉..............................適量

做法

1/ 糯米洗淨，用清水浸過面待 3 小時或以上，隔去水分。

2/ 在蒸籠或不銹鋼筲箕鋪上紗布，放入糯米鋪平，隔水蒸約 25 分鐘成糯米飯。

3/ 臘味料飛水約 3 分鐘，取出，臘肉再蒸 12 分鐘，全部切粒備用。

4/ 蝦米及冬菇分別用清水浸軟，冬菇去蒂，切幼粒。

5/ 大鍋內注入溫水（約 45℃），放入糯米飯用手弄散飯粒，洗去表面黏質，用筲箕瀝去水分。

6/ 鑊燒熱下生油 1 茶匙，放入臘味粒用小火炒香，下蝦米、冬菇同炒一會，加入糯米飯炒至熱透。

7/ 調味料混合，逐少加入糯米飯內炒勻，轉大火炒一會成糯米飯，待涼至暖。

8/ 紫菜片放於壽司蓆上，鋪上糯米飯，用包壽司手法捲起，捏成呈四方形的糯米卷，切 2 厘米厚件，蘸上蛋液，用中小火略煎香兩面，上碟享用。

Chef's Tips

- 糯米浸透數小時後不要大力洗擦，以免將米粒弄斷成碎粒。
- 糯米飯用紫菜包吃，滋味十足，可毋須煎香食用。
- 若沒時間炒製糯米飯，在外購買當然也可。
- 如糯米蒸熟後不即時炒製，可待糯米飯涼透後，用保鮮紙包好放於雪櫃，使用前用暖水浸至鬆軟即可。

蝦仁、蝦乾、蝦籽，三種口感、三款滋味

三蝦炒飯

材料

蝦仁……150 克
蝦乾……80 克
蝦籽……1 茶匙
菜心……6 條
甘筍……40 克
葱……1 棵
雞蛋液……2 個
白飯……4 碗
生油……1 湯匙

醃料

鹽……1/4 茶匙
粟粉……1 茶匙
麻油……半茶匙
胡椒粉……適量

調味料

鹽……2/3 茶匙
雞粉……1/3 茶匙

做法

1/ 蝦仁洗淨，吸乾水分，加入醃料拌勻待約半小時，切粒，飛水至熟，備用。

2/ 蝦乾洗淨，用清水浸軟，切粒。

3/ 菜心及甘筍洗淨，切薄片。

4/ 鑊燒熱下生油，用小火炒香蝦乾粒，倒入蛋液炒拌，放入蝦仁及白飯炒勻，轉中火炒至飯粒熱透。

5/ 加入調味料及菜粒快炒至熟，灑下蝦籽轉中大火快炒一會至香氣散發，熄火，灑入葱花略炒，上碟享用。

Chef's Tips

- 蝦籽不宜太早放入炒拌，容易過火散發焦燶氣味。
- 蝦乾不宜浸泡太長時間，令蝦乾鮮味減少；或可改用蝦米代替。

一口咬入，脆、嫩、鮮

滑蛋蝦仁煎米餅

材料

排米粉……250 克
蝦仁……200 克
雞蛋……1 個
葱花……20 克

蝦仁醃料

鹽……1/4 茶匙
胡椒粉……1/4 茶匙
麻油……1/3 茶匙
粟粉……1 茶匙

滑蛋芡汁料

鹽……1/3 茶匙
上湯……1 杯
粟粉……2 茶匙
胡椒粉……1/5 茶匙

滑蛋蝦仁煎米餅

做法

1/ 排米粉放入熱水煮至可弄散及稍軟即盛起，放於托盤上，用毛巾蓋密焗約 2 分鐘，揭去毛巾，用筷子撥散米粉（以免黏在一起），待涼。

2/ 蝦仁洗淨，用毛巾吸乾水分，放入醃料拌勻。

3/ 鑊燒熱下生油 2 湯匙，米粉弄成圓餅形放入鑊內，用中慢火煎香兩面至表面香脆，隔油，放於碟上剪成小件。

4/ 燒滾水放入蝦仁灼熟，盛起。

5/ 滑蛋芡汁料及蝦仁放入鑊煮熟至稠身，倒入蛋液拌勻成滑蛋芡，扒在米粉餅上享用。

Chef's Tips

- 排米粉不宜用滾水煮至腍身才盛起，炒米粉時容易折斷，甚至欠缺彈性質感。
- 米粉用毛巾蓋着，以餘溫令米粉保持彈性，而且不黏手。
- 煎米粉時火力不宜太猛，宜採用中慢火，如油分不足可逐少加入，但切勿過量，否則米粉吸附太多油分。

綿軟的粥底，是有味粥的靈魂

瑤柱白粥

此粥底可加入豬肉丸煮熟，
成為肉丸粥享用。

材料

白米………………………………1 杯

清水……………………約 2 公升

瑤柱………………………………8 粒

腐竹………1 片（約 60 克）

皮蛋………………………………1 個

生油…………1 茶匙（後下）

做法

1/ 瑤柱用清水浸過面至軟身（約 2 小時），搓散成絲。

2/ 白米洗淨，隔水，皮蛋去殼洗淨，放入白米內攪拌搓爛。

3/ 腐竹用清水浸半小時，隔水備用。

4/ 燒滾清水 2 公升，放入所有材料用大火煲滾，轉中慢火煲約 1 小時，加入 1 茶匙生油再煲約半小時，熄火即可。

Chef's Tips

- 白米加入皮蛋煲成的白粥，令白粥更綿軟，因皮蛋含鹼性令米粒軟化，但粥底不會太白。
- 白粥煲一小時後必須經常攪拌，避免黏底易焦燶。
- 如煲粥完成後，白粥太稠而水分不足，可加入適量滾水拌勻再煲一會。

青翠的米粒，伴有薑葱的清香

青葱瑤柱蟹籽炒飯

材料

瑤柱……5 粒
蟹籽……80 克
青葱……120 克
薑片……40 克
白米……250 克
蛋液……1 個
清水……200 毫升

調味料

鹽……1/3 茶匙
胡椒粉……1/5 茶匙
麻油……1/3 茶匙

做法

1/ 瑤柱用清水 80 毫升浸 2 小時至軟，放入鑊內蒸 45 分鐘，拆成絲，瑤柱水留用。

2/ 青葱洗淨、切段，與薑片放入攪拌機，加入清水 200 毫升攪碎，用密篩隔渣留用，薑葱水留作蒸飯用。

3/ 白米洗淨，隔去水分，放入電飯煲，加入薑葱水及瑤柱水煮成青葱飯（水與米的比例 1 比 1）。

4/ 鑊燒熱下生油 1 湯匙，放入蛋液略炒，下青葱飯用慢火炒匀，灑入調味料轉中火炒熱及米粒散開。

5/ 瑤柱、蟹籽及薑葱渣加入炒飯內拌匀，至飯粒乾身及帶香味即可上碟。

Chef's Tips

- 如薑葱碎渣太多，可減半份量放入炒飯內。
- 薑葱水最宜即用即做，否則薑葱水存放太久容易變壞，影響味道。

多滋味的鹹香湯丸，心滿又意足

釀豆卜鹹湯丸

湯丸皮材料

糯米粉……300克
粘米粉……35克

湯丸皮做法

1/ 粉料混合，取40克粉料與溫水20毫升搓勻成粉糰，放入沸水灼至半熟（約2分鐘），盛起備用。

2/ 其餘粉料加入溫水180毫升搓成麵糰，拌入半熟粉糰搓勻，成為湯圓皮。

Chef's Tips

- 如湯圓包好後不立即食用，宜放入雪櫃冷藏備用。
- 釀豆卜如想吃到香脆口感，可先用慢火煎香後才放入湯中。

餡料

五花腩肉……150克
榨菜粒……50克
蝦米……40克
香芹……40克
已浸發冬菇……4隻
薑米……20克

燜腩肉調味料

清水……3杯
鹽……1茶匙
雞粉……半茶匙
蠔油……2湯匙
八角……3粒
陳皮……1/3片
薑……3片
老抽……1湯匙

芡汁

燜腩肉湯……4湯匙
粟粉……1湯匙

釀豆卜材料

豆卜……20粒
鯪魚膠……400克

湯水配料

上湯……3杯
清水……3杯
鹽……半茶匙
娃娃菜絲……400克
釀豆卜……40粒

做法

1/ 煮熱燜腩肉調味料，放入五花腩肉用慢火燜約1小時至腍身。

2/ 所有餡料切成幼粒，放入鑊用芡汁拌勻成餡料，放入雪櫃冷藏備用。

3/ 豆卜切開兩邊，釀入鯪魚膠，備用。

4/ 取湯圓皮約25克，搓成圓形略壓扁，放入餡料1茶匙包實，搓成湯圓。

5/ 煮滾湯水配料，加入湯圓轉中火煮約6分鐘至浮起，上桌食用。

又香又脆的口感，是很大的誘惑

海鮮薄餅

材料

- 蝦仁……12 隻
- 帶子……8 粒
- 臘腸……半條
- 蝦米……40 克
- 甜菜脯……60 克
- 芫茜、葱粒……20 克

調味料

- 鹽……1/3 茶匙
- 粟粉……2 茶匙
- 麻油……1 茶匙

薄餅漿

- 糯米粉……120 克
- 低筋麵粉……90 克
- 雞蛋……1 個
- 清水……210 毫升
- 砂糖……1 湯匙
- 鹽……半茶匙
- 雞粉……1/3 茶匙
- 麻油……1 茶匙

做法

1/ 薄餅漿調好，備用。

2/ 蝦仁及帶子洗淨，吸乾水分，加入調味料拌勻，切幼粒備用。

3/ 臘腸切薄片；蝦米浸軟，切幼粒；甜菜脯切幼粒。

4/ 海鮮料灼熟；臘腸、蝦米及菜脯炒香，放入薄餅漿拌勻。

5/ 鑊燒熱下油 2 湯匙，倒入粉漿鋪平，放入海鮮料用中慢火煎香，灑上芫茜、葱粒，反轉用慢火煎香即可。

Chef's Tips

- 薄餅漿調配後如不立即使用，可用保鮮紙包好冷藏儲存。
- 煎薄餅時想有香脆的質感，在薄餅邊沿逐少加入油分，煎至表面香脆及金黃色即可。

CONTENTS

SEAFOOD

MEAT & POULTRY

SOYBEAN

VEGETABLE

CEREALS

Stir-fried Fish Fillets with Greens

Ingredients

- 300 g boned giant grouper
- 400 g Chinese kale
- 80 g Chinese celery
- 40 g preserved radish

Aromatics

- 1/2 tsp finely chopped garlic
- 10 g shredded ginger
- 10 g shredded carrot

Marinade

- 1/3 tsp salt
- 2 tsp corn flour
- 1 tbsp egg white (about half an egg)

Seasoning 1 (for blanching Chinese kale)

- 2 tsp salt
- 2 tsp sugar
- 2 tsp oil

Seasoning 2 (for stir-frying fish fillets)

- 1/3 tsp salt
- 1 tsp oyster sauce
- 1/3 tsp sesame oil
- ground white pepper
- 1 tsp corn flour
- 2 tsp water

Chef's Tips

- If the stems of Chinese kale are tough, you may skin the thicker part. Do not soak the stems in water for too long after both ends are cut crisscross; otherwise, they will shrink and curl too much and easily break during cooking.
- Run a small thin blade knife across the stems so that they will present beautifully after cooking.

Method *refer to p.17 for the steps

1. Remove the leaves of the Chinese kale and cut the stems into sections of about 5cm long. Cut a cross of 1.5cm deep on both ends of the stems with a small knife. Soak in water for about half an hour.
2. Thickly slice the fish fillets and cut into fine strips. Combine the marinade and mix well with the fish fillets.
3. Pick the leaves off the Chinese celery, cut the stems into small sections and then rinse. Cut the preserved radish into small sections and cut into fine strips. Soak in hot water to reduce saltiness.
4. Bring 2 cups of water in a wok to the boil. Blanch the Chinese kale with the seasoning 1 for about 2 minutes. Slightly blanch the Chinese celery and preserved radish, drain well.
5. Heat the wok, pour in around half a cup of oil and heat up over medium heat. Put in the fish fillets and deep-fry until done. Drain.
6. Sauté the aromatics in the wok until fragrant. Add the Chinese kale and slightly stir-fry. Put in the fish fillets and slowly pour in the mixed seasoning 2. Stir-fry with the other ingredients over medium heat and serve.

Stir-fried Squid with Pickled Mustard Greens, Fermented Black Bean and Chilli

Ingredients

- 300 g fresh squids
- 1 green bell pepper
- 1 red chilli
- 1/2 onion
- 200 g pickled mustard greens (refer to p.13 for method)

Aromatics

- 1/2 tsp finely chopped garlic
- 6 slices ginger
- 2 tsp fermented black beans

Seasoning

- 1/3 tsp salt
- 2 tsp oyster sauce
- 1 tsp dark soy sauce
- 1 tsp sugar
- 1 tsp corn flour
- 1 tbsp water
- 1 tsp Shaoxing wine (to be added later)

Chef's Tips

- It is easier to make a crisscross pattern by lightly scoring inside the body of the squid. Do not score too deep to avoid cutting it off.
- Cut away the tough part of the fins of the squid and cut the thin part into strips.
- Do not blanch the squid for too long. Overcooking will make it tough.
- The processed pickled mustard greens will have an additional sour, sweet flavour which is more delicious.

Method

*refer to p.20 for the steps

1. Seed the green bell pepper and red chilli, cut with the onion into triangles. Slice the pickled mustard greens diagonally and then stir-fry in a dry wok until dry and fragrant.
2. Skin the squid, cut open the body with scissors and rinse. Score the body to make a crisscross pattern and cut into small, long triangles.
3. Blanch the squid in boiling water and then drain.
4. Put 1 tbsp of oil in a wok and sauté the aromatics until fragrant. Add the green bell pepper, red chilli, onion and 2 tbsp of water, stir-fry over medium heat until done.
5. Put in the squids and pickled mustard greens and slightly stir-fry. Add the seasoning and give a good stir-fry over high heat. Finally sprinkle 1 tsp of Shaoxing wine and slightly stir-fry. Serve.

Fish Thick Soup with Pea Sprout

Ingredients

- 2 mandarin fishes (about 600 g each)
- 400 g pea sprout
- 1 light soft tofu
- 80 g carrot
- 50 g Jinhua ham
- 1 egg
- 3 cups stock (refer to p.10 for method)

Seasoning

- 1/2 tsp salt
- 1/4 tsp ground white pepper
- 1/3 tsp sesame oil

Corn flour solution

- 2 tbsp corn flour
- 4 tbsp water

Chef's Tips

- When you bone the fish, check it over for several times to avoid tiny bones embedded in the meat.
- Press water out of the pea sprout as much as you can; otherwise, the water will reduce the flavour of the soup.
- Cook the tofu in the final step to avoid breaking and overcooking.

Method

1. Rinse the mandarin fish, put on a plate, lay a couple of sliced ginger on top and steam for about 12 minutes, or until done. When it cools, bone it carefully and loosen the meat.
2. Rinse and finely chop the pea sprout. Put 1 tbsp of oil into a wok, add the pea sprout and 1 tsp of salt, stir-fry over medium heat until done. Press water out.
3. Finely slice the carrot. Cut the tofu into small cubes. Cover the Jinhua ham with water, steam for about 20 minutes and finely chop. Combine the Jinhua ham soup with the stock.
4. Bring the stock in the wok to the boil. Add the fish puree, pea sprout and carrot, and then bring to the boil. Put in the seasoning and tofu, stir in the corn flour solution and bring to the boil. Stir in the egg wash, transfer to the bowl and sprinkle the ham on top. Serve.

Pepper Crab with Basil

Ingredients

- 1 male mud crab
- 2 stalks basil
- 2 bunches fresh green
- peppercorns
- 1 tsp white peppercorns
- 1 red chilli
- 40 g butter

Aromatics

- 3 cloves garlic (sliced)
- 8 slices ginger
- 4 shallots (sliced)

Seasoning

- 1/2 tsp salt
- 1/2 tsp sugar
- 1 tbsp oyster sauce
- 1 tbsp Shaoxing wine
- 1 cup water

Chef's Tips

- It is not necessary to deep-fry the crab's shell as the roe inside will be overcooked. Just cook it with the other crab pieces with a lid on for 6 minutes.
- The time of cooking will vary according to the size of the crab. For a larger crab, it may need 2 to 3 minutes more to fully cook the claws. Additional water is also needed.

Method

1. Pick the leaves off the basil and keep the leaves. Bash the white peppercorns into coarse grains. Rinse the green peppercorns. Slice the red chilli.
2. Cut into the middle of the abdomen of the crab, and then pull open the shell. Remove the internal organs, rinse and chop into pieces. Crack the claws with a knife. Dust with corn flour.
3. Heat oil in a wok and deep-fry the crab until fully cooked. Drain.
4. Cook the butter until it dissolves. Stir-fry the aromatics and white peppercorns until fragrant. Put in the crab and slightly stir-fry. Pour in the seasoning and turn down the heat. Add the green peppercorns, put a lid on and then leave for about 6 minutes. Put in the basil leaves and red chilli, give a good stir-fry, and leave for 1 minute. Stir-fry until the sauce dries. Serve.

Steamed Grouper in Egg White and Coconut Water

Ingredients

- 500 g boned grouper
- 2 king coconuts
- 4 egg whites
- 1 sprig spring onion (shredded)
- 1 red chilli (shredded)

Seasoning 1 (egg white)

- 150 g egg white
- 150 ml king coconut water
- 1/3 tsp salt
- 1 tsp corn flour
- ground white pepper

Seasoning 2 (boned grouper)

- 1/3 tsp salt
- 1/4 tsp chicken bouillon powder
- 2 tsp corn flour
- 2 tsp egg white
- 1 tsp sesame oil
- 1/4 tsp ground white pepper
- 2 tsp Shaoxing wine
- 2 tsp finely chopped garlic
- 2 tsp finely diced ginger

Chef's Tips

- Dissolve the corn flour entirely in the coconut water before mixing it with the other ingredients of seasoning.
- Steam the grouper with the egg white mixture over low heat to avoid the egg white turning tough.

Method

1. Break open the king coconut, pour out the coconut water and keep the water. Remove the coconut flesh and finely dice. Combine the seasoning 1 and then mix well with the coconut flesh.
2. Cut the boned grouper into thick and rectangular pieces (about 4 x 2 cm), rinse and drain. Mix with the seasoning 2 and marinate for about 10 minutes.
3. Lay the grouper evenly on a deep dish and steam for about 4 minutes. Take out and discard the liquid. Add the egg white mixture and steam over low heat for 4 minutes, or until fully cooked. Sprinkle with the shredded ginger and red chilli. Serve.

King Razor Clams in Garlic Chilli Sauce

Ingredients

- 4 king razor clams
- 1 sprig spring onion (shredded)
- 1 stalk coriander
- shredded red chilli

Aromatics

- 8 cloves garlic (grated)
- 2 red chillies
- 30 g sliced ginger

Sauce

- 3 tbsp light soy sauce
- 1 tbsp oyster sauce
- 2 tsp sesame oil
- 1 tbsp Zhenjiang vinegar
- 1 tsp sugar
- 2 tbsp water

Chef's Tips

- Blanch the razor clams over high heat and control the time. Overcooking will make them shrink and tough.

Method

1. Put ice cubes into 4 cups of drinking water. Set aside.
2. Rinse the razor clams, blanch in boiling water for about 20 seconds, remove and chill in the ice water.
3. Wipe the razor clam meat dry, cut into 3 or 4 small sections. Put back on the shell and keep in a refrigerator.
4. Finely dice the aromatics. Put 1 tsp of oil into a wok and sauté the aromatics over low-medium heat until sweet-scented. Add the sauce and bring to the boil. Dish up.
5. Take out the razor clams from the refrigerator and pour the sauce on top. Finally sprinkle the shredded spring onion, red chilli and coriander on top. Serve.

Deep-fried Cuttlefish Cups

Chef's Tips

- To make a beautiful chrysanthemum, you do not require too much cuttlefish paste in one cup.
- The best width of the stripes is around 0.5 cm so that it will have more petals after deep-frying and look more beautiful.
- Allow the stripes to stretch out as far as possible during deep-frying. It will look more like a chrysanthemum.
- You could replace the cuttlefish paste with shrimp paste to get a similar result.

Ingredients

- 200 g cuttlefish paste (refer to p.11 for method)
- 6 square spring roll wrappers
- 30 g crab roe

Method

*refer to p.32 for the steps

1. Divide the cuttlefish paste into 12 small balls.
2. Fold up the spring roll wrapper. Cut into 2 rectangular sheets (10 x 20 cm each).
3. Fold up each rectangular sheet. Cut the folding part diagonally at 45 degrees into stripes with scissors. The depth is half that of the rectangular sheet while each stripe is about 0.5 cm to 1 cm wide.
4. Put the cuttlefish ball on the front part of the rectangular wrapper (the part without stripes). Roll inwards slowly to appear in the shape of a chrysanthemum. Seal with a little cuttlefish paste and fix with a toothpick.
5. Pour about 3 cups of oil into a wok, heat over medium heat until the temperature of oil reaches about 150°C. Put in the cuttlefish cups one by one, deep-fry until golden and fully cooked (about 3 minutes), set aside. Decorate with a little crab roe in the middle of the chrysanthemum. Serve.

Shrimps in Black Pepper Mint Sauce

Ingredients

- 12 fresh shrimps
- 200 g tri-colour bell peppers (seeded and cut into pieces)
- 50 g mint leaves

Black pepper mint sauce

- 2 tsp black peppercorns
- 2 tbsp mint jelly
- 10 fresh mint leaves (chopped)
- 1 tbsp oyster sauce
- 1/3 tsp salt
- 1 tsp sugar
- 3 tbsp water
- 1 tsp corn flour

Marinade

- 1/3 tsp salt
- 2 tsp corn flour
- 3 tsp egg wash
- 1/2 tsp sesame oil

Chef's Tips

- Mint sauce and mint jelly in glass jars are generally found in the market. Mint sauce is a bit sour and light while mint jelly is sweeter and thicker.
- Wipe dry on the mint leaves before deep-frying. Do not deep-fry over high heat because the leaves are very thin and easily burn.

Method

1. Heat the black pepper mint sauce. Set aside.
2. Rinse the mint leaves and wipe dry. Rinse the bell pepper and cut into triangles.
3. Shell the shrimps leaving the heads and tails. Slit the back of the shrimps to remove the intestine. Rinse and wipe dry the water with a dry cloth. Mix well with the marinade.
4. Put half a cup of oil into a wok and heat up to around 140°C. Deep-fry the mint leaves and remove. Deep-fry the shrimps until done and set aside.
5. Stir-fry the bell pepper in the wok until fragrant. Add the black pepper mint sauce and heat up. Put in the shrimps, give a good stir-fry and transfer to the plate. Sprinkle the crisp mint leaves on top. Serve.

Lemongrass Cuttlefish Balls

Ingredients

- 300 g frozen cuttlefish meat
- 6 peeled water chestnuts
- 6 sticks lemongrass
- 200 g breadcrumb

Seasoning

- 1/2 tsp salt
- 1/3 tsp chicken bouillon powder
- 3 tsp corn flour
- 1/2 tsp sesame oil
- 1/4 tsp ground white pepper

Method

*refer to p.38 for the steps

1. Rinse and dry the cuttlefish meat. Cut into small pieces and blend in a food processor. Rinse the water chestnuts and bash with a knife.
2. Put the cuttlefish meat into a big bowl and knead for a couple of times by hand. Stir in the same direction for about 1 minute, mix with the seasoning and stir until sticky. Take the meat and throw into the bowl for several times. Mix in the water chestnuts and stir into paste. Squeeze and shape into balls.
3. Rinse the lemongrass, cut away the end leaving the thicker part of around 10 cm long. Cut open in half. Slightly wet the hand and skewer the cuttlefish balls on the lemongrass. Coat with the breadcrumb.
4. Heat oil over medium heat and deep-fry the cuttlefish ball skewers until golden. Put on a plate and serve.

Chef's Tips

- Stir the cuttlefish meat in the same direction; otherwise, it will have a loose texture.
- If you are not ready to cook the cuttlefish paste, chill it in a refrigerator for over 2 hours to make it spongier.
- The lemongrass is cut open in half to let its beautiful smell penetrate into the fish balls.

Stir-fried Prawns on Asparagus

Chef's Tips

- The prawns' heads can be served as a side dish after deep-fried over medium heat and sprinkled with premium flavoured pepper, which is available in the supermarket.
- Better choose asparagus as thin as a chopstick.

Ingredients

- 6 large prawns
- 200 g fresh asparagus
- 80 g winter bamboo shoots
- 20 g Jinhua ham

Marinade

- 1/3 tsp salt
- 1 tsp corn flour
- 1/2 tsp sesame oil
- 1/4 tsp ground white pepper
- 2 tsp egg white

Aromatics

- 6 slices ginger
- 6 slices carrot
- 1 tsp finely chopped garlic

Seasoning

- 1/3 tsp salt
- 1 tsp oyster sauce
- 1/2 tsp corn flour
- 1 tbsp water

Method

*refer to p.41 for the steps

1. Cut away the head of the prawns; shell the body and keep the tail. Slightly slit the back of the prawns, rinse and wipe dry. Combine the marinade and mix well with the prawns.
2. Cover the Jinhua ham with water and steam for about 20 minutes. Remove and finely chopped.
3. Slice the bamboo shoots and blanch in boiling water for around 5 minutes to remove the unpleasant smell. Skin the asparagus and cut into two sections.
4. Make a hole respectively in the front of the prawn's back and tail. Pierce the sharp end of the asparagus through the holes on the front and the tail.
5. Pour 2 cups of water into a wok, add 1 tsp of salt, 1/2 tsp of sugar and 2 tsp of oil, and then heat up. Add the asparagus and prawns and blanch for about 2 minutes. Remove.
6. Put 1 tsp of oil into the wok and sauté the aromatics and bamboo shoots until fragrant. Pour in the asparagus and prawns and stir-fry for a while. Add the seasoning and quickly stir-fry over medium heat. Put on a plate, sprinkled with the Jinhua ham. Serve.

Soy Sauce Chicken with Perilla

Ingredients

- 1 chicken
- 4 eggs

Aromatics

- 4 star anise
- 20 g cinnamon
- 30 g galangal (sliced)
- 60 g skinned ginger (sliced)
- 6 shallots
- 2 sticks lemongrass (sectioned)
- 3 sprigs spring onion
- 1 whole dried tangerine peel (soaked in water to soften; scrape off the pith)

Seasoning

- 500 ml light soy sauce
- 200 ml Indonesian sweet soy sauce
- 10 cups water
- 100 ml Shaoxing wine
- 600 g slab sugar
- 30 g salt
- 30 g chicken bouillon powder
- 2 stalks basil
- 10 dried perilla

Chef's Tips

- Discard all the ingredients in the soy sauce mixture after the chicken is removed. Bring the soy sauce mixture to the boil again to avoid spoiling. Store it when it cools down.
- Submerge the whole chicken in the soy sauce mixture to heat evenly and let the flavour permeate the chicken.
- The key to make smooth chicken is to leave it in the soy sauce mixture for half an hour after the heat is turned off.

Method

1. In a large pot, put in 2 tbsp of oil. Stir-fry the aromatics over low-medium heat until fragrant. Pour in the seasoning and bring to the boil. Turn to low heat and simmer for about 1 hour.
2. Rinse the chicken, put into the soy sauce mixture and bring to the boil. Turn down the heat, simmer for about 5 minutes. Turn off the heat and leave for about 30 minutes.
3. Cook the eggs in boiling water until done, remove the shells, soak into the soy sauce mixture with the chicken for half an hour.
4. Take out the chicken, chop into pieces. Cut open the eggs in half, put on the side and then pour the soy sauce mixture on top. Serve.

Stir-fried Beef Tenderloin and Assorted Mushrooms in Black Pepper Sauce

Ingredients

- 300 g beef tenderloin
- 6 fresh shiitake mushrooms
- 10 button mushrooms
- 100 g shaggy ink cap
- 100 g tri-colour bell peppers
- 2 tbsp black pepper sauce (refer to p.15 for method)

Aromatics

- 1 tsp finely chopped garlic
- 10 g sliced ginger

Marinade

- 2 tsp light soy sauce
- 2 tsp corn flour
- 2 tbsp water
- 1 tsp Shaoxing wine

Seasoning

- 2 tsp oyster sauce
- 1/2 tsp sugar
- 1 tsp corn flour
- 1 tbsp water
- 1/4 tsp ground black pepper

Chef's Tips

- If you choose a tougher part of beef, add a little fruit juice like lemon juice or papaya juice into the marinade. The fruit acids will help soften the meat texture.

Method

1. Rinse and dice the beef tenderloin, mix with the marinade and rest for about 1 hour.
2. Rinse and dice all the mushrooms, blanch in boiling water for a while and drain. Rinse the bell peppers and cut into triangles.
3. Heat a wok, add 1 tbsp of oil and fry the beef tenderloin over low-medium heat until done and fragrant. Put in the aromatics, stir-fry over medium heat until sweet-scented. Add the black pepper sauce, bell peppers and mushrooms, stir-fry for a moment.
4. Mix the seasoning, pour into the wok, give a quick stir-fry over high heat, serve.

Crisp Pork Spareribs with Pistachio

Ingredients

- 400 g pork spareribs
- 250 g shelled pistachio
- 6 cloves garlic
- 40 g skinned ginger
- 80 g low-gluten flour

Marinade

- 1/2 tsp salt
- 1/2 tsp chicken bouillon powder
- 2 tsp Shaoxing wine
- 2 tbsp egg wash
- 3 tsp corn flour
- 2 tbsp water

Seasoning for salad dressing

- 150 ml salad dressing
- 1 tbsp peanut butter
- 3 tbsp hot drinking water
- 1 tbsp condensed milk
- 1 tsp fresh lemon juice

Chef's Tips

- Do not dip the spareribs into the salad dressing right after they are deep-fried; otherwise the salad dressing will melt.
- It is not easy to dissolve peanut butter in cold water. Dissolve it in hot water and then mix well with the other seasoning. Do not mix them together in one go.
- Keep any unused seasoned salad dressing in a refrigerator for use in 1 or 2 days.

Method

1. Chop the spareribs into around 2-inch long sections, rinse and drain.
2. Finely dice the garlic and ginger, mix well with the marinade. Combine with the spareribs and marinate for about 1 hour.
3. Crush the pistachio into fine dices.
4. Stir the peanut butter in hot water until smooth. Add the other seasoning, mix well and chill in a refrigerator.
5. Put half a wok of oil into the wok, heat up over medium heat. Coat the spareribs with the low-gluten flour, deep-fry until fully cooked and golden crisp, drain well.
6. When the spareribs cool down a little bit, dip into the salad dressing. Coat with finely chopped pistachio and serve.

Braised Pigeon with Roast Pork and Imitation Shark's Fin

Ingredients

- 1 pigeon
- 100 g imitation shark's fin
- 30 g Jinhua ham
- 2 dried shiitake mushrooms
- 100 g roast pork (cut into small pieces)
- 6 cloves garlic
- 4 slices ginger

Seasoning

- 1/2 tsp salt
- 2 tbsp oyster sauce
- 1 tsp dark soy sauce
- 1/2 tsp sugar
- 1 tbsp Chu Hou sauce
- 1/4 tsp ground white pepper
- 2 tsp Shaoxing wine
- 2.5 cups water

Chef's Tips

- If you are not prepare to take this dish right after cooking, just simmer it for about half an hour, turn off the heat and keep it in a refrigerator when it cools. Steam it for about 15 minutes and thicken the sauce with corn flour solution before serving.

Method

*refer to p.52 for the steps

1. Gut the pigeon, rinse and wipe dry. Soak the imitation shark's fin in water to soften and drain.
2. Soak the dried shiitake mushrooms in water until soft, remove the stalks, blanch in boiling water for about 15 minutes, or until tender. Cut into shreds and set aside.
3. Cover the Jinhua ham with about 100 ml of water and steam for about 20 minutes. Finely cut into strips and reserve the ham stock.
4. Put the ham stock into a wok. Add the mushrooms, Jinhua ham and imitation shark's fin, mix well and heat up. Slightly thicken the stock with corn flour solution.
5. Fill the pigeon cavity in full with all the above ingredients, seal with a toothpick, and then colour the skin with dark soy sauce. Heat the wok, put in 1 tbsp of oil, slightly fry the skin over medium heat until it colours, transfer to a small casserole.
6. Put 1 tbsp of oil into the wok, fry the garlic over low heat until golden and fragrant. Add the ginger and roast pork, stir-fry until aromatic. Put in the seasoning and heat up, pour into the casserole to cover the pigeon.
7. Put a lid on the casserole and simmer the pigeon for about 40 minutes, or until tender. Pour out the sauce and thicken with corn flour solution. Remove the toothpick, cut open the pigeon chest, add the sauce and bring to the boil. Serve.

Curry Beef Omelet

Ingredients

- 4 eggs
- 160 g minced beef
- 80 g onion (finely chopped)
- 8 button mushrooms
- 1 stalk coriander (finely chopped)
- 2 tsp curry paste
- 10 g butter

Marinade

- 1/4 tsp salt
- 1 tsp corn flour
- 2 tsp egg wash
- 1 tsp water
- ground white pepper

Seasoning

- 1/3 tsp salt
- 1/3 tsp chicken bouillon powder
- 1/3 tsp sesame oil
- 1 tsp corn flour
- 2 tsp water

Chef's Tips

- Drain the stir-fried beef and other ingredients before mixing them with the egg wash. It will be difficult to set if the egg wash is watery.

Method

1. Mix the minced beef with the marinade and rest for about half an hour.
2. Blanch the button mushrooms in boiling water, finely dice and set aside.
3. Combine the corn flour of the seasoning with 2 tsp of water, and then mix with the other ingredients of the seasoning. Add 3 eggs and whisk together.
4. Heat a wok and put in 1 tbsp of oil. Stir-fry the minced beef over low-medium heat until done, set aside.
5. Heat the wok, add the butter and curry paste. Stir-fry the onion, button mushrooms and beef until fragrant. Put into the egg wash with the coriander and mix well.
6. Heat the wok and put in oil. Swirl the wok to let the oil spread evenly, and then pour out leaving about 1 tsp of the oil in the wok. Turn to low heat, put in the egg mixture, fry both sides until done, looking like a round cake shape, and then cut into triangles.
7. Whisk an egg and dip the omelet into the egg wash. Fry over low heat until the surface is golden crisp. Serve.

Double-steamed Meatballs with Peking Cabbage

Ingredients

- 8 stalks Peking cabbage
- 500 g pork belly
- 100 g fresh yam
- 40 g Jinhua ham
- 4 cups stock
  (refer to p.10 for method)
- 40 g sliced ginger

Seasoning

- 2/3 tsp salt
- 1/2 tsp chicken bouillon powder
- 80 ml water
- 1 egg
- 1 tsp Shaoxing wine
- 30 g diced ginger
- 1/4 tsp ground white pepper
- 2 tsp corn flour

Method

1. Skin the pork belly, rinse and wipe dry. Finely slice, cut into thin strips, and then finely dice. Slightly chop to make pork mince.
2. Add the water of the seasoning to the pork bit by bit. Stir in the same direction until sticky, mix in the other ingredients of the seasoning, and then stir until gluey.
3. Skin the yam, rinse and finely dice. Add yam into the pork paste and shape into balls.
4. Bring water to the boil, turn to low heat and put in the balls to shape. Transfer lightly to a tureen.
5. Finely dice the Jinhua ham. Trim the big leaves of the Peking cabbage, cut open in half and rinse. Blanch the Jinhua ham and Peking cabbage in boiling water, and put into the tureen.
6. Bring 4 cups of stock with ginger to the boil. Pour into the tureen to cover the ingredients, double-steam for about 1 hour and 30 minutes. Serve.

Chef's Tips

- The best pork belly consists of 30% fat and 70% lean meat without tendons.
- Chill the skinned and rinsed pork belly in a refrigerator to make it firmer, so that it is easier to finely slice and dice it.
- The meatballs should be made in the same size and in the weight of about 60 to 80 g each.
- If you dislike fresh yam, replace it with water chestnuts.

Stir-fried Spicy Pork in Mixed Sauce

Ingredients

- 300 g pork belly
- 120 g yam beans
- 100 g preserved radish
- 100 g green, red bell pepper (diced)
- 3 five-spice dried tofu
- 60 g dried prawns
- 50 g diced onion
- 50 g deep-fried peanuts

Aromatics

- finely chopped garlic
- sliced ginger
- diced spring onion

Seasoning 1
(for stewing pork belly)

- 1/2 tsp salt
- 1/2 tsp chicken bouillon powder
- 2 tbsp oyster sauce
- 2 tsp dark soy sauce
- 15 g rock sugar
- 3 star anise
- 3 large ginger slices
- 2 spring onions
- 3 cups water

Seasoning 2

- 2 tsp Hoi Sin sauce
- 3 tsp chilli bean sauce
- 1 tsp Chu Hou sauce
- 1 tbsp oyster sauce
- 1/3 tsp salt
- 1 tsp sugar
- 2 tsp Zhenjiang vinegar
- 1 tsp dark soy sauce
- 2 tbsp water
- 1 tsp corn flour

Method

1. Blanch the pork belly in boiling water and rinse. Bring the seasoning 1 to the boil and add the pork belly. Cook over low heat for about 45 minutes and dice.
2. Soak the dried prawns in water until soft, coarsely dice with the other ingredients.
3. Soak the yam beans, five-spice dried tofu and preserved radish in hot water until hot, and then drain.
4. Combine the ingredients of seasoning 2 together. Heat a wok, add 2 tsp of oil, sauté the aromatics until fragrant, add the yam beans, preserved radish, bell peppers, dried tofu, dried prawns and onion, stir-fry until aromatic.
5. Put in the pork belly and seasoning 2, stir-fry quickly over high heat until even. Sprinkle with the deep-fried peanuts, serve.

Chef's Tips

- To save the stewing process, replace the pork belly with char siu (BBQ pork) comprising half fat and half lean meat.

Steamed Taro Pork in Lotus Leaf

Ingredients

- 600 g pork belly
- 500 g taro
- 1 tbsp finely chopped garlic
- 1 dried lotus leaf

Seasoning

- 200 ml water
- 2 large red fermented bean curd
- 1 tbsp Chu Hou sauce
- 1/3 tsp salt
- 3 tbsp sugar
- 3 tbsp oyster sauce
- 2 tbsp Shaoxing wine

Method

*refer to p.62 for the steps

1. Rinse the pork belly. Blanch in boiling water for about 30 minutes and rinse. Colour the surface with dark soy sauce and fry the skin.
2. Skin the taro and cut into thick rectangular slices (about 3 x 5cm). Deep-fry over medium heat until fragrant. Set aside.
3. Soak the dried lotus leaf in water until soft. Blanch it in boiling water for a while to remove the unpleasant smell, wipe dry and lay on a big bowl.
4. Cut the pork belly into thick slices in the size similar to that of taro.
5. Put 1 tbsp of oil into a wok, sauté the garlic and seasoning until aromatic, set aside.
6. Alternate the pork belly and taro in the bowl with taro leaf. Pour in the seasoning and wrap up. Steam over medium heat for about 2 hours and 30 minutes. Serve.

Chef's Tips

- Cut the taro into thick slices. A thin piece will easily break in the steaming process.
- Instead of deep-frying, you can pan-fry both sides of the taro to firm.
- The small pork pieces can be arranged in the gap between the taro and pork to make it more compact.
- You can defer serving the steamed pork for one day. Keep it in a refrigerator and steam it to reheat before serving. It will be more delicious.

Chef's Tips

For chicken wings in smaller size, reduce the baking time to avoid overbaking and burning.

Baked Chicken Mid-joint Wings with Stuffed Chestnuts

Ingredients

- 10 chicken mid-joint wings
- 10 large chestnuts
- 10 sliced bacon
- 4 tbsp honey

Marinade

- 3 tbsp Hoi Sin sauce
- 1 tbsp Chu Hou sauce
- 2 tbsp oyster sauce
- 1/3 tsp salt
- 1 tsp chicken bouillon powder
- 4 tbsp sugar
- 1 tbsp sesame oil
- 1 tbsp Shaoxing wine
- 2 tbsp finely chopped garlic
- 20 g sliced ginger
- 1 egg

Method

*refer to p.65 for the steps

1. Cook the chestnuts in boiling water until done, remove the shells and skin. Set aside.
2. Bone the chicken wings, combine the marinade and mix well with the chicken wings. Leave to marinate for about half an hour.
3. Soak the bacon in warm water for a while to reduce saltiness, wipe dry and set aside.
4. Stuff the chestnut into the chicken wing, roll with the bacon, and then fix with a toothpick.
5. Preheat an oven, put the chicken wings on a baking tray, bake at 220°C for about 5 minutes, and then bake at 150°C for about 8 minutes. Set aside.
6. Brush honey on the chicken wings, and bake again at 200°C for about 2 minutes or until they give a hint of charred fragrance. Serve.

Steamed Pork Belly with Chilli and Belachan

Ingredients

- 300 g pork belly
- 1 red chilli
- 30 g skinned ginger
- 80 g dried prawns
- 1 sprig spring onion

Seasoning

- 1 tbsp belachan (refer to p.14 for method)
- 1/4 tsp salt
- 2 tsp sugar
- 1 tsp corn flour
- 1/2 tsp sesame oil

Method

1. Rinse the pork belly and finely slice.
2. Shred the red chilli, ginger and spring onion.
3. Soak the dried prawns in water until soft, drain and then put into the pork. Add the ginger, red chilli and seasoning, mix well and lay on a plate.
4. Steam over medium heat for about 12 minutes. Finally sprinkle with the spring onion. Serve.

Chef's Tips

- Slice the pork belly with skin on to make it chewier after steamed.
- After the pork belly is rinsed and dried, chill it in a refrigerator. It is easier to slice the pork belly when it is firm.

Braised Beef Shank in Chinese Spice Marinade with Preserved Mustard Tuber

Ingredients

- 500 g beef shank
- 2 preserved mustard tuber (zha cai)
- 1 red chilli

Seasoning for mustard tuber

- 2 tbsp sugar
- 1 tbsp sesame oil
- 2 tsp chilli bean sauce

Aromatics of Chinese marinade

- 10 g liquorice
- 10 g red yeast rice
- 3 nutmegs
- 6 g sand ginger
- 5 g Sichuan peppercorns
- 4 star anise
- 1/3 dried tangerine peel
- 6 cloves garlic
- 40 g sliced ginger
- 2 sprig spring onion
- 3 stalks coriander

Seasoning

- 5 cups water
- 100 g rock sugar
- 60 ml light soy sauce
- 1/2 tsp salt
- 2 tsp chicken bouillon powder
- 1 tsp five-spice powder
- 40 ml Shaoxing wine

Chef's Tips

- If the beef shank is too big, cut it into two pieces before blanching. It helps the flavour seep into the beef in the cooking process.

Method

1. Rinse and dice the preserved mustard tuber. Soak in warm water to reduce saltiness and drain. Mix with the seasoning for mustard tuber.
2. In a large pot, add 1 tbsp of oil, stir-fry the aromatics over low heat until fragrant. Pour in the seasoning and bring to the boil. Cook over low heat for about 1 hour.
3. Blanch the beef shank in boiling water for about 5 minutes and rinse. Put into the Chinese spice marinade and bring to the boil. Turn down the heat, simmer for about 1 hour, turn off the heat and leave until cool. Set aside.
4. Finely dice the red chilli; cut the beef shank into cubes, mix well with the preserved mustard tuber. Sprinkle with a little Chinese spice marinade. Serve.

Tofu Nuggets with Chilli Pepper

Ingredients

- 1 firm tofu
- 4 salted egg yolks
- 80 g butter

Salted water solution

- 1 cup warm water
- 1 tsp fine salt

Chilli pepper batter

- 200 g low-gluten flour
- 20 g baking powder
- 2 tsp Japanese chilli pepper (Shichimi)
- 1/3 tsp salt
- 200 ml water
- 1 tbsp oil (to be added later)

Method

1. Combine the ingredients of the batter together, leave for about half an hour, add 1 tbsp of oil, and then mix well. Set aside.
2. Steam the salted egg yolks for about 15 minutes, or until done. Crush to pieces and set aside.
3. Cut the tofu into 1-inch cubes, soak in salted water for about 20 minutes. Wipe dry with a dry cloth.
4. Heat oil in a wok, turn to medium heat, dip the tofu in the batter and deep-fry until golden and crisp. Set aside.
5. Dissolve the butter in the wok over low heat, add the salted egg yolks and give a good stir-fry. Put in the crisp tofu, mix well and serve.

Chef's Tips

- By soaking in salted water, protein in the tofu will be firm, hence releasing less water.
- Do not use the batter right after it is mixed. Let it rest for at least half an hour for fermentation to make it crisp after deep-frying.
- Dissolve the butter over low heat before stir-frying with the salted egg yolk; otherwise it will hardly be runny like oozing sand.

Stir-fried Egg White Soya Milk with Shrimps

Ingredients

- 300 g shelled shrimps
- 8 eggs
- 150 ml light soya milk
- 120 g corn flakes
- 10 g finely chopped Jinhua ham (refer to p.9 for method)

Seasoning for egg whites

- 1/3 tsp salt
- 1/3 tsp chicken bouillon powder
- 2 tsp corn flour
- 3 tsp water

Marinade

- 1/2 tsp salt
- 2 tsp corn flour
- 1 tsp sesame oil
- ground white pepper

Method

1. Separate the 8 egg yolks from the egg whites. For the seasoning of egg whites, dissolve the corn flour in water, and then combine with the other ingredients. Put into the egg whites and mix well.
2. Clean the shrimps with corn flour and water, wipe dry with a dry cloth, mix well with the marinade.
3. Pour the soya milk into the seasoned egg whites and mix well.
4. Blanch the shrimps in boiling water, deep-fry in oil over medium heat. Set aside.
5. Pour the soya milk mixture into a wok, stir-fry over low heat until 80% done. Put in the shrimps and stir-fry together until fully cooked.
6. Lay part of the corn flakes on a dish. Put the cooked soya milk mixture on top. Sprinkle with the Jinhua ham, and put the rest corn flakes on the side. Serve.

Chef's Tips

- The commonly used fresh milk is replaced with light soya milk so that people who are not suitable for taking fresh milk can also try. Soya milk has less water and more starch, which will set easily during cooking.
- To make creamy egg white soya milk, the wok for stir-frying needs to be hot enough, but the heat should not be too high. You may turn off the heat at the final stage and stir-fry the milk mixture until done.
- The rest egg yolks can be used for decoration. Combine the egg yolks with 3 tbsp of water, slowly heat some oil in a non-stick pan and fry the egg wash into a thin sheet. Shred it and put on the side of the dish.

Claypot Tofu with Shrimp Roe and Fried Gluten Balls

Ingredients

- 10 fried gluten balls
- 6 dried shiitake mushrooms
- 2 cloth-wrapped tofu
- 8 stalks Shanghai bok choy
- 2 tsp shrimp roe

Seasoning

- 1/3 tsp salt
- 1 tbsp oyster sauce
- 1 tsp dark soy sauce
- 1/2 cup water

Chef's Tips

- You may choose shrimp roe in jars. If use shrimp roe in bulk, first stir-fry it in a dry wok until fragrant; keep in a small jar when cools for future use.
- You may fry every side of the tofu over medium heat. Do not stir the tofu for many times while stewing to avoid breaking.
- Deep-fried cloth-wrapped tofu has a silky texture while the firm tofu tastes tough.

Method

1. Blanch the fried gluten balls in boiling water to soften, chill in cold water and rinse to remove greasiness.
2. Soak the dried shiitake mushrooms in water until soft, rinse and cut into pieces.
3. Cut each tofu into 6 small pieces, deep-fry in hot oil (about 170°C) until golden. Set aside.
4. Put 1 tsp of oil into a wok, stir-fry 1/2 tsp of finely chopped garlic and a couple slices of ginger until fragrant. Put in the seasoning and heat up, add all the ingredients and simmer for about 2 minutes. Thicken the sauce with 2 tbsp of corn flour solution, transfer into a clay pot.
5. Arrange the Shanghai bok choy on the side, bring to the boil over medium heat, finally sprinkle with the shrimp roe. Serve.

Mud Carp Tofu Skin Rolls

Ingredients

- 1 tofu skin
- 300 g mud carp paste
- 1 preserved pork sausage
- 1 preserved liver sausage
- 40 g dried shrimps
- 10 g diced spring onion
- 10 g finely chopped coriander

Batter

- 200 g low-gluten flour
- 20 g baking powder
- 1/3 tsp salt
- 220 ml water
- 1 tbsp oil (to be added later)

Method

*refer to p.80 for the steps

1. Mix the batter and then rest for half an hour.
2. Soak the dried shrimps in water until soft, and then finely chop. Mix well with the mud carp paste, add the spring onion and coriander, give a good stir. Throw into a bowl for a couple of times.
3. Blanch the preserved pork sausage and liver sausage in boiling water for about 1 minute, wipe them dry and cut open in half. Cut the tofu skin into 2 rectangular sheets (the width is the same as the length of the sausage).
4. Spread the mud carp paste evenly on the tofu skin, arrange the pork sausage and liver sausage on top, roll up to wrap the mud carp paste tightly. Wrap in aluminum foil and steam for 8 minutes, or until done. Take out to cool.
5. Heat oil in a wok over medium heat. Dip the mud carp rolls in the batter, deep-fry until golden and crisp. Cut into pieces and serve.

Chef's Tips

- It is more convenient to buy seasoned mud carp paste in the market, and add other ingredients on your own.
- Wrap the mud carp rolls in aluminum foil or cover them with a cloth before steaming. This is to prevent the tofu skin from getting wet by dripping of condensed water.

Shredded Tofu, Tofu Skin and Mixed Vegetables in Thick Soup

Ingredients

- 1 cloth-wrapped tofu
- 1 fresh tofu skin
- 200 g winter melon
- 200 g celtuce (stem lettuce)
- 120 g carrot
- finely chopped Jinhua ham
- 2 cups stock
- 1 egg white

Seasoning

- 1/2 tsp salt
- 1/4 tsp ground white pepper
- 1/3 tsp sesame oil

Corn flour solution

- 2 tbsp corn flour
- 4 tbsp water

Chef's Tips

- Shredding tofu is not so difficult as you imagine. It is easier to shred by using a chopping knife with a large and sharp blade.

Method

*refer to p.83 for the steps

1. Finely slice the tofu, cut into shreds and soak in warm drinking water.
2. Finely slice the other ingredients, cut into fine shreds. Blanch in boiling water with 1 tsp of salt until done. Set aside.
3. Pour the stock into a wok and heat up. Add the seasoning and mix well. Put in the ingredients (except the tofu) and heat up. Thicken the soup with the corn flour solution.
4. When the soup turns thicken, put the drained tofu into the soup, slightly stir and mix in the egg white. When it forms a pattern, turn off the heat at once and sprinkle the Jinhua ham on top. Serve.

Salted Egg Yolk and Pumpkin Tofu Pudding

Ingredients

- 2 bowls light tofu pudding
- 4 salted egg yolks
- 300 g pumpkin
- 30 g Jinhua ham
- 1 sprig spring onion (diced)
- 1/2 cup water

Seasoning

- 1/3 tsp salt
- 1/3 tsp sesame oil

Method

1. Rinse the Jinhua ham, cover with half a cup of water and steam for about half an hour. Drain, keep the ham water and finely chop the ham.
2. Steam the salted egg yolks for about 8 minutes, crush into pieces. Dice the pumpkin and steam for about 5 minutes. Set aside.
3. Pour the ham water, pumpkin and salted egg yolks into a wok. Add the seasoning and heat up, thicken the sauce with 2 tbsp of corn flour solution.
4. Put the tofu pudding into a bowl, ladle the pumpkin and salted egg yolk over. Sprinkle with the ham and spring onion. Serve while warm.

Chef's Tips

- If you want to reheat the tofu pudding, use low heat to steam it for 5 minutes. Do not use high heat because it will overcook and taste tough.

Stir-fried Bitter Gourd with Dried Tofu and Olive-pickled Leaf Mustard

Ingredients

- 150 g minced pork
- 3 pieces five-spice dried tofu
- 300 g bitter gourd
- 2 tbsp olive-pickled leaf mustard

Aromatics

- 1/2 tsp finely chopped garlic
- 1/2 tsp diced ginger
- 1/2 tsp diced spring onion
- 1 tsp fermented black beans

Marinade

- 1/4 tsp salt
- 2 tsp corn flour
- 4 tsp water

Thickening glaze

- 1/3 tsp salt
- 2 tsp oyster sauce
- 1 tsp dark soy sauce
- 1 tsp sugar
- 3 tsp water
- 1 tsp corn flour

Chef's Tips

- To reduce bitterness, scrape the white pith off the blanched bitter gourd and then rinse well.
- You may choose long-shaped bitter gourds for stir-fries. For stews, pick the short and cone-like ones.

Method

1. Combine the marinade and mix well with the minced pork.
2. Cut open the bitter gourd in half and remove the seeds. Blanch in boiling water for about 3 minutes, chill in cold water, wipe dry and cut into pieces.
3. Dice the dried tofu, soak in boiling water until hot, drain and set aside.
4. Heat a wok, put in 1 tbsp of oil and stir-fry the pork until fully cooked. Set aside.
5. Put 1 tsp of oil in the wok, sauté the aromatics and olive-pickled leaf mustard until fragrant. Put in the other ingredients, stir-fry until sweet-scented. Finally add the thickening glaze and give a good stir-fry. Serve.

Fried Tofu with Shrimp Paste and Dried Oysters

Ingredients
(for making 16 pieces)

- 1 cloth-wrapped tofu
- 180 g shrimp paste
  (refer to * for ingredients)
- 10 dried oysters
- 4 dried shiitake mushrooms
- 1 egg
- 1 stalk coriander
- 20 g finely chopped Jinhua ham

Seasoning 1
(for stewing dried oysters and shiitake mushrooms)

- 4 slices ginger
- 2 sprigs spring onion
- 1/2 tsp salt
- 1 tsp oyster sauce
- 1 tsp sugar
- 1 tbsp Shaoxing wine
- 2 cups water

Seasoning 2

- 1/2 tsp salt
- 1 tbsp oyster sauce
- 1 egg yolk
- 3 tbsp corn flour
- 1/4 tsp ground white pepper
- 1/2 tsp sesame oil

*To prepare shrimp paste

Ingredients

- 180 g shelled shrimps
- 2 tsp corn flour
- 1/3 tsp salt
- 1/3 tsp sesame oil

Method

1. Rub the shelled shrimps with 1 tbsp of corn flour, rinse and wipe dry with a dry cloth.
2. Bash the shrimps with a knife and finely chop with the back of the knife.
3. In a big bowl, stir the shrimp puree in the same direction until sticky. Add the salt and sesame oil, stir and throw into the bowl repeatedly until gluey.

Method

*refer to p.90 for the steps

1. Soak the dried shiitake mushrooms and oysters in water until soft. Remove the stalks of the mushrooms and rinse. Remove sand from the oysters and rinse.
2. Heat seasoning 1, add the shiitake mushrooms and oysters. Simmer for half an hour, drain and finely dice.
3. Rinse the coriander, pick off the leaves and set aside. Finely chop the stems.
4. Press water out of the tofu, put into a bowl, add seasoning 2 and mix well. Mix in the shiitake mushrooms and oysters. Add the shrimp paste and mix well.
5. Grease porcelain soup spoons with oil, scoop a level spoonful of the paste. Lay the coriander leaves and Jinhua ham on top. Steam over medium heat for about 8 minutes. Take out to cool. With a small knife, remove the paste in the shape of a pear-liked from the spoon.
6. Dip the tofu in the egg wash, fry both sides over medium heat until fragrant, serve.

Chef's Tips

- The seasoning for tofu contains egg yolk and corn flour, which make the paste to firm.
- The paste will swell right after steaming. Let it cool down a little bit. It will be easily removed from the spoon.
- If you are not ready to use the tofu, soak it in water and keep it in a refrigerator to avoid any unpleasant smell.
- Any kinds of ingredients like bacon can be added to make the dish extraordinary. Be creative!

Chinese Celery and Bean Curd Stick Salad

Ingredients

- 250 g celery
- 2 deep-fried bean curd sticks
- 2 pieces dried wood ear
- 1 red chilli

Dressing

- 1/2 tsp salt
- 2 tbsp oyster sauce
- 2 tsp chilli bean sauce
- 1 tsp sugar
- 2 tsp sesame oil
- 1 tsp finely chopped garlic

Method

1. Soak the deep-fried bean curd sticks and wood ear in warm water until soft. Cut the bean curd sticks into small sections. Shred the wood ear.
2. Peel the hard skin off the celery and rinse. Cut into small sections and then cut into fine strips.
3. Bring 2 cups of water to the boil, add 2 tsp of salt and 1 tsp of oil. Blanch the celery, bean curd sticks and wood ear until done. Cool in cold drinking water, drain and refrigerate for 1 hour.
4. Mix the dressing, pour into the ingredients, mix well and serve.

Chef's Tips

- Do not toss the ingredients in the dressing too early. It contains salt which, after refrigeration, will release water reducing the flavour. Just mix them right before serving.

Stuffed Lotus Root and Pork Belly Stew

Ingredients
- 600 g pork belly
- 400 g lotus root
- 150 g shelled mung beans

Aromatics
- 10 shallots
- 4 large ginger slices
- 4 sprigs spring onion (sectioned)
- 6 star anise
- 1/3 dried tangerine peel

Seasoning
- 1 cup light soy sauce
- 1 cup Shaoxing wine (Hua Diao)
- 2 tbsp oyster sauce
- 2 tbsp dark soy sauce
- 120 g rock sugar
- 1 tsp salt
- 1 tsp chicken bouillon powder
- 2 tbsp Chu Hou sauce
- 12 cups water

Method *refer to p.97 for the steps

1. Cook the pork belly in hot water for about 20 minutes, take out and rinse in cold water.
2. Soak the mung beans in water for about 1 hour, cook in hot water for around 4 minutes, drain well.
3. Skin the lotus root, rinse and cut open the tip. Fill the holes with mung beans, cover the tip and fix with a bamboo skewer.
4. In a pot, add a little oil to sauté the aromatics until fragrant. Put in the seasoning, heat up, add the pork belly and lotus root, stew over low heat for about 1.5 hours. Take out the pork belly and chill in a refrigerator when cool. Stew the lotus root again for 30 minutes, or until tender.
5. Cut the pork belly and lotus root into chunks, arrange on a deep dish, pour in the cooked sauce and steam for about 15 minutes. Serve.

Chef's Tips

- Use a thin chopstick to help pack the mung beans into the lotus root. Do not stuff too many beans because they will swell after cooking to crack the lotus root. Allow some space among the beans.
- It requires your patience to stuff the mung beans. With a chopstick, press the mung beans into the lotus root, turning the chopstick slowly. Stuff again when there is room.
- If you find the pork belly and lotus root are not tender enough when cutting them, steam again until they are soft.

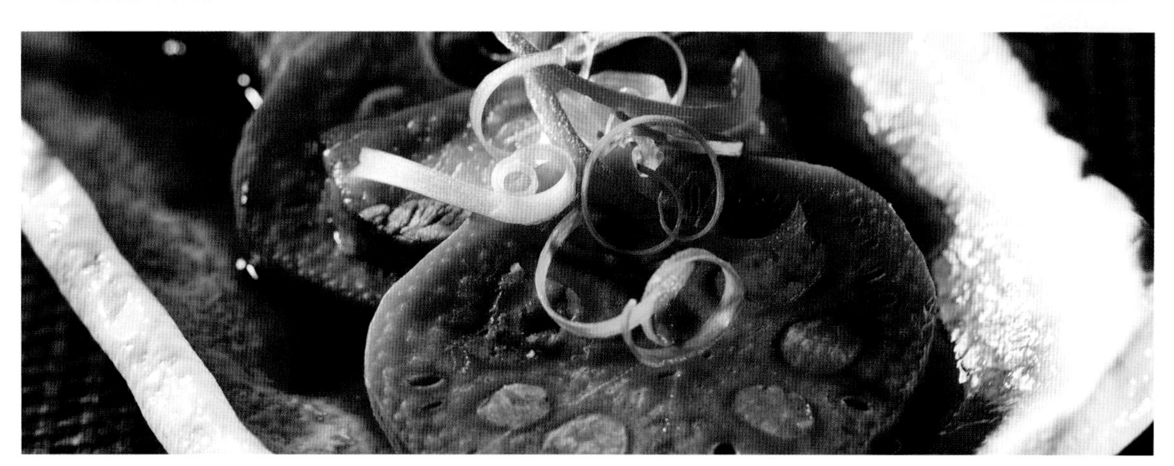

Stir-fried Fresh Yam with Salted Pork

Ingredients

- 300 g pork belly
- 200 g fresh yam
- 100 g Chinese celery
- 4 fresh shiitake mushrooms

Marinade for pork belly

- 2 tbsp salt
- 1 tsp sugar
- 2 tbsp Shaoxing wine
- 1 tsp five-spice powder
- 1/3 tsp ground white pepper

Aromatics

- 1/2 tsp finely chopped garlic
- 6 slices ginger

Seasoning

- 1/3 tsp salt
- 1 tsp oyster sauce
- 1 tsp corn flour
- 2 tsp water

Chef's Tips

- If you dislike the skin of the pork belly, remove it before slicing the pork.
- Put on gloves before skinning the fresh yam to avoid itchy hands.

Method

*refer to p.100 for the steps

1. Cut the pork belly into long pieces and rinse. Combine the marinade, mix well with the pork belly, put into a zipper bag and chill in a refrigerator for 1 day.
2. Take out the pork belly, rinse to remove salt. Blanch in boiling water for about 15 minutes, take out and finely slice.
3. Skin the yam, rinse and cut into slices. Pick the leaves off the Chinese celery, and cut the stems into small sections. Slice the shiitake mushrooms.
4. Put the yam, Chinese celery, shiitake mushrooms into a wok, add 2 tsp of oil, 1/2 tsp of salt and 3 tbsp of water, stir-fry over medium heat until fully cooked. Set aside.
5. Heat 2 tsp of oil, stir-fry the pork belly until done. Add the aromatics and sauté over medium heat until fragrant. Put in the other ingredients, give a good stir-fry, mix the seasoning and sprinkle on the wok, stir-fry evenly over medium heat. Serve.

Baby Cucumbers and Fresh Abalones in Sake

Ingredients

- 6 fresh abalones
- 2 baby cucumbers

Seasoning

- 2 star anise
- 20 g dried wolfberries
- 1 cup water
- 1/2 tsp salt
- 1/2 tsp sugar
- 20 g dried bonito flakes
- 100 ml sake (to be added at last)

Method

1. Heat the seasoning over low heat until hot, leave to cool, mix well with the sake. Set aside.
2. Cut open the baby cucumbers in half and then cut into small sections. Mix well with 2 tsp of salt and leave to marinate for about 3 minutes. Rinse to remove salt.
3. Rub the shells of the fresh abalones and rinse. Steam for 6 minutes, or until fully cooked, soak in ice water. When cool, remove the shells and internal organs, and wash thoroughly.
4. Soak the cucumbers and abalones in the seasoning, refrigerate for 1 day. Take out on the next day, cut into slices and serve.

Chef's Tips

- Use baby cucumbers to make this dish, because big and large ones can hardly absorb the flavour.
- Do not cook or heat the sake to let the wine flavour evaporate.

Steamed White Radish Meatballs

Ingredients

- 300 g pork ball filling (refer to p.12 for method)
- 500 g white radish
- 1 lotus leaf
- 20 g finely chopped Jinhua ham
- diced spring onion

Chef's Tips

- If fresh lotus leaf is not available, use the dried one instead, but it needs to be softened before blanching. Soak it in water to make it soft first.
- If you want to add some colour to the dish, mix the white radish with shredded carrot. Prepare the carrot in the same way as you do with the white radish. Marinate the carrot with table salt, it will turn soft quickly while steaming.

Method

1. Squeeze and shape the pork ball filling into 12 small balls.
2. Skin the white radish, finely slice and cut into shreds. Mix well with 2 tsp of table salt and leave to marinate for about 3 minutes. When they turn soft, rinse to remove salt. Drain well.
3. Blanch the lotus leaf in boiling water, wipe it dry and lay on a plate.
4. Coat the pork balls with the white radish, knead lightly to firm and put on the lotus leaf. Steam for about 10 minutes. Sprinkle with the Jinhua ham and spring onion. Serve.

*refer to p.104 for the steps

Bamboo Fungus Rolls Stuffed with Pea Sprout

Ingredients

- 10 dried bamboo fungus
- 400 g pea sprout
- 200 g Enokitake mushrooms

Seasoning

- 1/2 cup stock
  (refer to p.10 for method)
- 1/3 tsp salt
- 2 tsp corn flour
- ground white pepper

*refer to p.108 for the steps

Chef's Tips

- Use warm light salt water to rehydrate the bamboo fungus. When they turn soft, rinse them repeatedly to remove granules and look white.
- Wild bamboo fungus is yellowish in colour with a delicate fragrance, and is sold at a higher price. Artificially propagated bamboo fungus is whiter with a hint of sulphur smell which needs to be removed by blanching in boiling water. It is cheaper.
- The meshed part of the bamboo fungus cut out can be reserved to make soup.
- Apart from pea sprout, other leaf vegetables like spinach can also be used as filling.

Method

1. Soak the bamboo fungus in light salt water for about 2 hours to soften. Cut away the tip and the meshed part with scissors, leaving the stalk for cooking.
2. Slightly blanch the bamboo fungus in boiling water to remove the unpleasant smell, drain well. Simmer with 1 cup of stock to make flavourful.
3. Cut away the root of the Enokitake mushrooms, cut into small sections and make loose.
4. Rinse the pea sprout and finely chop. Put 1 tbsp of oil into a wok, stir-fry the pea sprout with 1 tsp of salt over medium heat until cooked, press water out.
5. Drain the bamboo fungus, stuff with the pea sprout in full, arrange on a plate and steam for about 6 minutes.
6. Deep-fry the Enokitake mushrooms over medium heat (about 150°C) until crunchy, drain.
7. Take out the bamboo fungus rolls and discard the water. Heat the seasoning, stir into thickening glaze, pour over the bamboo fungus rolls. Lastly sprinkle with the deep-fried Enokitake mushrooms. Serve.

Taro Puffs with Crab Meat

Ingredients

- 900 g taro puree (refer to * for ingredients)
- 100 g crab meat
- 1 egg (whisked)

Seasoning

- 1 tsp salt
- 2 tsp sugar
- 1 tsp chicken bouillon powder
- 1/2 tsp five-spice powder
- 1 tsp sesame oil

Thickening glaze for crab meat

- 1/3 tsp salt
- 1/2 cup stock (refer to p.10 for method)
- 3 tsp corn flour
- 1/5 tsp ground white pepper

Method

1. Divide the taro dough into around 10 small portions, knead into circular pieces, deep-fry over medium heat (about 150°C) until they puff. Turn to high heat (about 180°C) and deep-fry until golden crisp.
2. Bring the stock and crab meat to the boil. Mix the thickening glaze and add to the stock, stir until thicken. Slowly stir in the egg wash, cook until done. Pour into the centre of the taro puffs. Serve.

*To prepare taro puree

Ingredients

- 600 g taro
- 100 g tang flour
- 80 g vegetable shortening

Method

1. Skin the taro, cut into small pieces, steam for 20 minutes or until tender. Mash into puree when it cools down a little bit.
2. Mix tang flour with 100 ml of boiling water, knead into a half done dough. Add the taro puree, vegetable shortening and seasoning, mix well to be taro puree.

Chef's Tips

- Discard any lumpy taro when kneading the dough to make a smooth texture.
- The taro dough will not shape while deep-frying if the temperature of oil is too low. Adjust the heat to make crumbly puffs.

Scrambled Eggs with Mung Bean Sprouts, Dried Scallops and Imitation Shark's Fin

Ingredients

- 150 g imitation shark's fin
- 200 g mung bean sprouts
- 5 dried scallops
- 4 eggs
- 1 stalk coriander

Seasoning

- 1/3 tsp salt
- 1/4 tsp ground white pepper
- 1/2 tsp sesame oil

Method

1. Cover the dried scallops with water and soak for about 2 hours. Steam with the water for 45 minutes, tear into shreds and keep the soaking scallop water.
2. Blanch the imitation shark's fin in boiling water for a while, take out and soak in the soaking scallop water to absorb flavour.
3. Heat a wok, put in 1 tbsp of oil, quickly stir-fry the mung bean sprouts over medium heat. Add 1 tsp of salt and stir-fry until 60% done, drain.
4. Whisk the eggs with the seasoning. Heat the wok, add 1 tbsp of oil, put in the egg wash and stir-fry over low-medium heat until half done. Add the imitation shark's fin and dried scallops, turn to medium heat and stir-fry until dry and fragrant.
5. Finally put in the mung bean sprouts, quickly stir-fry until even, transfer to a plate. Sprinkle the coriander on top. Serve hot.

*refer to p.112 for the steps

Chef's Tips

- When stir-frying the mung bean sprouts the first time, it is no need to add water and fully cook them; or they will release water when stir-fried with the eggs and will not be crunchy.
- The characteristic of this dish is to scramble the eggs until dry and broken into granules, looking like the sweet osmanthus flowers, and they are also beautifully scented.

Miso Radish Rings

Ingredients

- 10 white radish rings
- 2 salted egg yolks
- diced spring onion

Thickening glaze

- 1 tbsp corn flour
- 2 tbsp water

Seasoning for stewing radish

- 2 tbsp miso
- 20 g dried bonito flakes
- 3 star anise
- 2 tsp dark soy sauce
- 4 slices ginger
- 1/2 tsp salt
- 20 g rock sugar
- 3 cups water

Chef's Tips

- Thick radish is not suitable for making this dish. Choose thin and straight ones.
- If you want the radish rings appear in the same shape, you may use a stainless steel round mould to do it.

Method

1. Skin the white radish, cut into a ring shape and rinse.
2. Steam the salted egg yolks until done, put into a small stainless steel strainer and crush into granules with a spoon.
3. Bring the seasoning to the boil, put in the radish rings and cook over low heat for about 40 minutes.
4. Put the radish rings on a plate, strain the soup. Combine the thickening glaze with 1 cup of the soup, pour over the radish rings. Finally sprinkle the salted egg yolks and spring onion on top. Serve.

Chef's Tips

- Do not use piping hot stock to rehydrate the mung bean vermicelli. It will overly swell and become too soft when cooking.
- Keep less skin on the angled luffa as the skin is a bit tough with a strong grassy smell.

Garlic Steamed Angled Luffa with Mung Bean Vermicelli

Ingredients

- 1 angled luffa (about 400 g)
- 1 small pack mung bean vermicelli
- 80 g garlic
- 2 large slices ginger

Seasoning

- 2 tbsp oil
- 1 tsp sesame oil
- 1/3 tsp salt
- 1/3 tsp sugar
- 1 tbsp oyster sauce
- 1 tsp light soy sauce

Method

1. Pour 2 cups of water into a pot, add 1/2 tsp of salt and 1 tsp of oyster sauce, and then heat up. When cool, soak in the mung bean vermicelli.
2. Skin the angled luffa alternately, cut open to make 4 strips. Remove the pith and cut into sections, rinse and then drain.
3. Finely chop the garlic. Finely dice the ginger.
4. Heat a wok, put in the oil of the seasoning, stir-fry the ginger and garlic over low heat until fragrant, add the other seasoning and mix well.
5. Drain the mung bean vermicelli, put on a plate. Mix the garlic seasoning with the angled luffa, arrange on the mung bean vermicelli and steam for about 8 minutes. Serve.

Clay Pot Rice with Dried Prawns and Pork Back Ribs

Chef's Tips

- Pay attention to the water of rice when putting in the pork back ribs. Do not wait until the rice is too dry because it will hardly make the pork to cook through.
- When the heat is turned down, move the clay pot around repeatedly to avoid the heat staying in the same position to burn the rice.
- If the rice is too dry and not fully cooked, drizzle some drinking water over the rice and cook it over low heat until done.

Ingredients

- 250 g pork back ribs
- 60 g dried prawns
- 6 dried shiitake mushrooms
- 6 red dates
- 300 g white rice

Marinade

- 1/3 tsp salt
- 2 tsp oyster sauce
- 2 tsp corn flour
- 4 tsp water
- 6 slices ginger
- 2 tsp oil

Sauce

- 3 tbsp light soy sauce
- 2 tsp dark soy sauce
- 2 tsp sugar
- 1 tsp oil
- 1 tsp sesame oil
- 3 tbsp water
- ground white pepper
- 1 stalk coriander

Method

*refer to p.121 for the steps

1. Rinse the dried prawns and soak in water until soft.
2. Soak the dried shiitake mushrooms in water until soft, remove the stalks and cut into slices. Rinse and core the red dates.
3. Chop the pork back ribs into small pieces and rinse. Mix with the marinade, shiitake mushrooms, red dates and dried prawns, and leave for about half an hour.
4. Rinse the white rice, and put into a clay pot. Add 300 ml of water and cook over low-medium heat until the rice water reduces. Add the pork back ribs at once, lay flat on the rice, cook over low heat with a lid on for about 12 minutes, or until the rice are done. Turn off the heat and leave for about 6 minutes.
5. Mix the sauce and bring to the boil, discard the coriander. Drizzle the clay pot rice with the sauce.

Fried Glutinous Rice Rolls with Preserved Sausages and Dried Shrimps

Ingredients

- 300 g glutinous rice
- 1 preserved sausage
- 1 preserved liver sausage
- 1/2 preserved pork
- 6 dried shiitake mushrooms
- 50 g dried shrimps
- 1 stalk coriander (finely chopped)
- 1 sprig spring onion (diced)
- 2 eggs (whisked)
- 4 laver sheets

Seasoning

- 1 tbsp light soy sauce
- 1 tbsp oyster sauce
- 1 tsp dark soy sauce
- 1/2 tsp chicken bouillon powder
- 1 tbsp drinking water
- 1 tsp sesame oil
- ground white pepper

Method

*refer to p.124 for the steps

1. Rinse the glutinous rice. Cover with water and soak for at least 3 hours, drain.
2. Lay a piece of gauze cloth on a bamboo steamer or stainless steel strainer. Put in the glutinous rice and make them flatten. Steam for about 25 minutes.
3. Blanch all the preserved meat in boiling water for about 3 minutes, remove. Steam the preserved pork again for 12 minutes, and then dice all.
4. Soak dried shrimps and shiitake mushrooms respectively in water until soft, remove the stalks of the mushrooms and finely dice.
5. In a large pot, pour in warm water (about 45°C). Put in the glutinous rice and separate the rice with hands, wash away the sticky substance on the surface. Strain the water off the rice.
6. Heat a wok and put in 1 tsp of oil. Stir-fry the preserved meat over low heat until aromatic. Add the dried shrimps and shiitake mushrooms, stir-fry for a while, put in the glutinous rice and stir-fry until hot through.
7. Mix the seasoning and add to the glutinous rice bit by bit while stir-frying. Turn to high heat, stir-fry for a while and leave to cool.
8. Put the laver sheet onto the sushi mat, lay the glutinous rice on top, roll up and knead into a square glutinous rice roll. Cut into 2-cm thick pieces, dip into the egg wash, and slightly fry both sides over low-medium heat until fragrant. Serve.

Chef's Tips

- Do not rub the glutinous rice heavily after soaking it for a couple of hours to avoid breaking.
- You may choose not to fry the glutinous rice roll as it is delicious by just eating the laver wrapped in the glutinous rice.
- If you have no time to make the glutinous rice, you can certainly buy the ready-made outside.
- If the steamed rice is not ready to be cooked, wrap it in cling film when it is cool and then refrigerate it. Before cooking, soak the rice in warm water to make it soft.

Stir-fried Rice with Shrimps, Dried Prawns and Shrimp Roe

Ingredients

- 150 g shelled shrimps
- 80 g dried prawns
- 1 tsp shrimp roe
- 6 stalks flowering Chinese cabbage
- 40 g carrot
- 1 sprig spring onion
- 2 eggs (whisked)
- 4 bowls cooked rice
- 1 tbsp oil

Marinade

- 1/4 tsp salt
- 1 tsp corn flour
- 1/2 tsp sesame oil
- ground white pepper

Seasoning

- 2/3 tsp salt
- 1/3 tsp chicken bouillon powder

Chef's Tips

- Do not put in the shrimp roe for stir-frying too early. It will easily burn with a scorched smell.
- Soak the dried prawns in water for just a while to keep their flavour. They can be replaced with dried shrimps.

Method

1. Rinse the shelled shrimps, wipe dry. Mix with the marinade and leave for about half an hour. Cut into dices and blanch in boiling water until done.
2. Rinse the dried prawns, soak in water to soft and cut into dices.
3. Rinse the flowering Chinese cabbage and carrot, and then finely slice.
4. Heat a wok and add some oil. Stir-fry the dried prawns over low heat until aromatic, add the egg wash and stir-fry. Put in the shrimps and cooked rice, give a good stir-fry, turn to medium heat, stir-fry until the rice is fully hot.
5. Add the seasoning, flowering Chinese cabbage and carrot, give a quick stir-fry until done. Sprinkle with the shrimp roe, turn to medium-high heat and stir-fry quickly until fragrant. Turn off the heat, sprinkle the diced spring onion on top, stir-fry slightly and serve.

Rice Vermicelli Cake with Egg and Shrimps

Ingredients

- 250 g rice vermicelli in dried form
- 200 g shelled shrimps
- 1 egg
- 20 g diced spring onion

Marinade

- 1/4 tsp salt
- 1/4 tsp ground white pepper
- 1/3 tsp sesame oil
- 1 tsp corn flour

Thickening glaze

- 1/3 tsp salt
- 1 cup stock (refer to p.10 for method)
- 2 tsp corn flour
- 1/5 tsp ground white pepper

Chef's Tips

- Do not cook the rice vermicelli to totally soft. It will easily break when stir-frying and will not give a spongy texture.
- The rice vermicelli is covered with a towel so that the remaining heat will keep it spongy and will not stick to the hands.
- Do not use high heat to fry the rice vermicelli. Cook it over low-medium heat, and if there is not enough oil during frying, add in some oil little by little. Do not put in too much oil as it will make the rice vermicelli greasy.

Method

*refer to p.130 for the steps

1. Cook the rice vermicelli in hot water until they are a bit soft and can be loosened. Take out immediately, transfer into a tray, cover entirely with a towel for about 2 minutes. Remove the towel, loosen with chopsticks to avoid sticking together, leave to cool.
2. Rinse the shelled shrimps, wipe dry with a towel, mix well with the marinade.
3. Heat a wok, add 2 tbsp of oil. Shape the rice vermicelli into a round cake, put into the wok and fry over low-medium heat until both sides are crisp and fragrant. Drain and cut into small pieces.
4. Blanch the shrimps in boiling water until done and set aside.
5. Put the shrimps and thickening glaze into the wok and cook until done and thicken. Mix in the egg wash and put on top of the rice vermicelli cake. Serve.

Rice Congee with Dried Scallops

Chef's Tips

- The rice congee is softer by cooking the rice with preserved duck's egg. The alkaline preserved duck's egg helps soften the rice but the congee base will not look totally white.
- Stir the rice congee frequently after cooking it for 1 hour to avoid sticking to the pot and getting burnt.
- If the rice congee is done but appears too thick, mix it with some boiling water and cook it again for a while.
- You can add pork balls into this congee base and cook until the pork is done as meatball congee.

Ingredients

- 1 cup white rice (the cup that comes up rice cookers)
- 2 litres water
- 8 dried scallops
- 1 dried tofu sheet (about 60 g)
- 1 preserved duck's egg
- 1 tsp oil (to be added later)

Method

*refer to p.133 for the steps

1. Cover the dried scallops with water, soak until soft (about 2 hours). Rub into shreds.
2. Rinse the white rice and drain. Shell the preserved duck's egg, rinse and mix with the rice and crush to pieces.
3. Soak the dried tofu sheet in water for half an hour, drain.
4. Bring 2 litres of water to the boil. Put in all the ingredients, bring to the boil over high heat, turn to low-medium heat and cook for about 1 hour. Add 1 tsp of oil and cook again for about half an hour. Turn off the heat and serve.

Stir-fried Rice with Spring Onion, Dried Scallops and Crab Roe

Ingredients

- 5 dried scallops
- 80 g crab roe
- 120 g spring onion
- 40 g sliced ginger
- 250 g white rice
- 1 egg wash
- 200 ml water

Seasoning

- 1/3 tsp salt
- 1/5 tsp ground white pepper
- 1/3 tsp sesame oil

Method

*refer to p.136 for the steps

1. Soak the dried scallops in 80 ml of water for 2 hours to soften. Steam for 45 minutes and tear into shreds. Keep the soaking scallop water.
2. Rinse the spring onion and cut into sections. Blend spring onion, ginger and 200 ml of water in a food processor, strain the ginger and spring onion for later use, and keep the sauce for steaming rice.
3. Rinse and drain the white rice, put into a rice cooker. Add the ginger spring onion sauce and soaking scallop water (the proportion of water to rice is 1:1), cook until done.
4. Heat a wok and put in 1 tbsp of oil. Stir-fry the egg wash for a while, add the rice and stir-fry over low heat until even. Sprinkle with the seasoning, turn to medium heat and stir-fry until the rice separates.
5. Add the dried scallops, crab roe and blended ginger and spring onion to the rice, stir-fry until the rice is dry and fragrant, serve.

Chef's Tips

- If too much blended ginger and spring onion remains, put only half portion of it into the rice.
- Cook the ginger and spring onion sauce right after it is made. It will spoil if leaving it for too long and will not taste good.

Savory Glutinous Rice Balls with Stuffed Tofu Puffs

Ingredients of glutinous rice ball skin

- 300 g glutinous rice flour
- 35 g rice flour

Ingredients of filling

- 150 g pork belly
- 50 g diced preserved mustard tuber (zha cai)
- 40 g dried shrimps
- 40 g Chinese celery
- 4 rehydrated shiitake mushrooms
- 20 g diced ginger

Seasoning for stewing pork belly

- 3 cups water
- 1 tsp salt
- 1/2 tsp chicken bouillon powder
- 2 tbsp oyster sauce
- 3 star anise
- 1/3 dried tangerine peel
- 3 slices ginger
- 1 tbsp dark soy sauce

Thickening glaze

- 4 tbsp pork belly soup
- 1 tbsp corn flour

Ingredients of stuffed tofu puff

- 20 deep-fried tofu puffs
- 400 g mud carp paste

Ingredients of soup

- 3 cups stock
- 3 cups water
- 1/2 tsp salt
- 400 g shredded baby cabbage
- 40 stuffed tofu puffs

Chef's Tips

- If you are not ready to cook the glutinous rice balls just made, keep them in a refrigerator.
- If you want to eat crisp stuffed tofu puffs, fry them over low heat before putting them into the soup.

Method for glutinous rice ball skin

1. Combine the glutinous rice flour with the rice flour. Mix 40 g of the combined flour with 20 ml of warm water, knead into dough. Blanch in boiling water until half cooked (about 2 minutes), Set aside.
2. Mix the rest combined flour with 180 ml of warm water, knead into dough. Mix in the half cooked dough to form glutinous rice ball skin.

Method

1. Bring the seasoning to a boil, put in the pork belly and simmer over low heat for about 1 hour, or until tender.
2. Finely dice all the ingredients of filling. Put them into a wok, mix with the thickening glaze to form the filling, refrigerate for later use.
3. Cut the tofu puffs in half and stuff with the mud carp paste.
4. Take about 25 g of the glutinous rice ball skin, knead into a round shape. Slightly flatten and wrap 1 tsp of the filling, knead to form glutinous rice balls.
5. Bring the ingredients of soup to the boil, add the glutinous rice balls, turn to medium heat and cook for about 6 minutes, or until they float. Serve.

Seafood Pizza

Chef's Tips

- If you are not prepare to use the batter right after it is mixed, wrap it in cling film and keep it in a refrigerator.
- If you want to fry a crisp pizza, add some oil little by little around the edge of the pizza and fry until the surface is golden crisp.

Ingredients

- 12 shelled shrimps
- 8 scallops
- 1/2 preserved sausage
- 40 g dried shrimps
- 60 g sweet preserved radish
- 20 g coriander and diced spring onion

Seasoning

- 1/3 tsp salt
- 2 tsp corn flour
- 1 tsp sesame oil

Batter

- 120 g glutinous rice flour
- 90 g low-gluten flour
- 1 egg
- 210 ml water
- 1 tbsp sugar
- 1/2 tsp salt
- 1/3 tsp chicken bouillon powder
- 1 tsp sesame oil

Method

*refer to p.140 for the steps

1. Mix the batter.
2. Rinse the shelled shrimps and scallops, wipe dry. Mix with the seasoning and finely dice.
3. Finely slice the preserved sausage. Soak the dried shrimps in water until soft and finely dice. Cut the sweet preserved radish into fine cubes.
4. Blanch the seafood in boiling water until cooked. Stir-fry the preserved sausage, dried shrimps and preserved radish until aromatic, put into the batter and mix well.
5. Heat a wok and add 2 tbsp of oil. Pour in the batter and flatten. Add the seafood and fry over low-medium heat until fragrant. Sprinkle with the coriander and spring onion, turn over and fry over low heat until fragrant. Serve.

簡材料 • 滋味餸

Gourmet Cooking with Simple Food

作者 Author
廖教賢 Alvin Liu

策劃/編輯 Project Editor
Karen Kan

攝影 Photographer
Imagine Union

美術設計 Design
Nora Chung

出版者 Publisher
Forms Kitchen

香港鰂魚涌英皇道1065號
東達中心1305室
Room 1305, Eastern Centre, 1065 King's Road,
Quarry Bay, Hong Kong.
電話 Tel: 2564 7511
傳真 Fax: 2565 5539
電郵 Email: info@wanlibk.com
網址 Web Site: http://www.wanlibk.com
http://www.facebook.com/wanlibk

萬里機構

萬里 Facebook

萬里 Instagram

發行者 Distributor
香港聯合書刊物流有限公司
SUP Publishing Logistics (HK) Ltd.
香港新界大埔汀麗路36號
中華商務印刷大厦3字樓
3/F., C&C Building, 36 Ting Lai Road,
Tai Po, N.T., Hong Kong
電話 Tel: 2150 2100
傳真 Fax: 2407 3062
電郵 Email: info@suplogistics.com.hk

承印者 Printer
百樂門印刷有限公司
Paramount Printing Co., Ltd.

出版日期 Publishing Date
二零一七年七月第一次印刷
First print in July 2017

版權所有 · 不准翻印
All right reserved.
Copyright©2017 Wan Li Book Co. Ltd

Published in Hong Kong by Forms Kitchen,
a division of Wan Li Book Company Limited.
ISBN 978-962-14-6466-8